T0342530

Geology and the Pioneers of Earth Science

DEDICATION

In memory of Phoebe.

...here lay the hound, old Argos.
But the moment he sensed Odysseus standing by
he thumped his tail, nuzzling low, and his ears dropped,
though he had no strength to drag himself an inch
toward his master. Odysseus glanced to the side
and flicked away a tear...

Homer, *The Odyssey*; transl. Robert Fagles, Penguin Classics (1997); Book 17, lines 329–334

Also by Mike Leeder:
GeoBritannica, Geological landscapes and the British peoples
With Joy Lawlor (2016)
Measures for Measure, Geology and the Industrial Revolution (2020)

Other Liverpool University Press (formerly Dunedin) books that may be of interest:
Breakthroughs in Geology, Ideas that transformed Earth Science
Graham Park (2019)
The Abyss of Time, A study in geological time and Earth history
Paul Lyle (2015)

Geology and the Pioneers of Earth Science

Mike Leeder

LIVERPOOL UNIVERSITY PRESS

First published 2024 by
Liverpool University Press
4 Cambridge Street
Liverpool
L69 7ZU

British Library Cataloguing-in-Publication data
A British Library CIP record is available

ISBN 978-1-78046-106-9 (hardback)
ISBN 978-1-78046-682-8 (ePub)
ISBN 978-1-78046-683-5 (PDF)

Typeset by Carnegie Book Production, Lancaster
Printed and bound by Hussar (Poland).

Contents

Cover Illustration

Hot Rock – Brenda Hartill. Reproduced by kind permission of the artist: © Brenda
Hartill (www.brendahartill.com)
The artist lived in New Zealand when young – school field trips with Mr Rose, an
inspirational geography teacher, spent climbing the volcanoes of North Island's central
plateau inspired a lifetime love of landscapes and how they formed.

Acknowledgements

This project has been at the back of my mind for decades – I thank Anthony Kinahan of Dunedin Academic Press for reading the eventual proposal and for his encouragement to proceed. Health issues have slowed and occasionally halted progress – my partner, Hilary Joy Lawlor, and my brother and sister-in-law, Andy and Jan, gave support and help when most needed.

The creative seeds required for writing such a book were sowed initially by the exceptionally wide syllabi in geology that were made available to me as a student at the English universities of Durham and Reading. These were followed by the example set by my colleagues as a young staff member at the University of Leeds, the first earth sciences school in Britain, founded by the legendary W.Q. Kennedy at the time of his retirement in 1967.

Those seeds were encouraged to sprout by the wise advice and example of two noteworthy individuals. The first was Brian Funnell, fellow 'old boy' of the City of Norwich School, pioneering Quaternary micropalaeontologist and co-founder of the first School of Environmental Sciences in Britain – UEA Norwich. In late 1968, Brian, with his usual frankness, advised me that I should reconsider and broaden my initial ideas for Ph.D. research if I was to achieve anything remotely original.

The second was one of the pioneers featured in this book, the only one whom I had the good fortune to meet. This was Ralph Bagnold, who, after correspondence concerning tests I had proposed for his 1966 sediment transport theory, invited me down from Leeds to discuss things in person at his home in Kent. I walked that morning in the late spring of 1975 to his house from the local railway station through the sunlit birdsong of hedged Kentish lanes. On the scientific front he encouraged me in his modest but firm way to delve deeper into the physics of turbulence as affected by entrained solid particles. On the hospitality front I remember two things: Mrs Bagnold's Yorkshire pudding and roast potatoes with rich meaty gravy (I think it was venison) at lunchtime and the lift he gave me in the afternoon back to catch my train in his air-cooled Renault – 'Superior engine to anything British', or words to that effect, he chuckled.

Writing the book has made me acutely aware of the precious gifts of selfless kindness and modesty that permeated the lives of both Funnell and Bagnold. Also, after delving deeply into the lives of my choice of earth science pioneers it seems to me now that scientific originality often bestows graciousness, as well it should.

I thank the following guardians of records at the: British Geological Survey (Luke McDonald, Bob McIntosh, Andrew Morrison, Rachel Talbot); Caltech Archives and Special Collections, Pasadena, California (Elisa Piccio); Carnegie Science, Washington D.C. (Shaun Hardy); the NHS GGC Archives, Glasgow, Scotland (Laura Stevens); the National Portrait Gallery, London; the Scottish National Portrait Gallery (Alejandro Basterrechea) and the University of Arizona Press (Julia Balestracci) for their help in obtaining copyrighted images and/or for other information.

Thanks go to Steve Hencher of the University of Leeds for initial help with the Karl Terzaghi chapter. I am especially indebted to Kate Johnston of Jedburgh, Scotland for consulting and making copious notes from hitherto unpublished documents in the Glasgow Hospital Archives that form the personal sections of Chapter 9 concerning the life and times of Ernest Masson Anderson.

I also thank friends and ex-colleagues Jim Best and Julian Andrews dating back to my University of Leeds and University of East Anglia days, and to Graham Park and John Dewey for going through the final manuscript, offering encouragement and suggestions. The factual errors that doubtless remain are *mea culpa*. Thanks here to Anne Morton, Kerrie Moncur and all at Liverpool University Press and Carnegie Book Production involved in the copy-editing, layout, composition and production of this book.

Aims

I have given a great deal of thought as to whom I should be addressing in this book and to what degree the book should be made 'geo-friendly', enabling it to be read with enjoyment (perhaps with a little help from Wikipedia) by non-geoscientists. Of course, I wanted to reach out to my fellow earth scientists interested in the history of their subject, but also to involve those many laypersons who are curious about what is perhaps the most profound, relevant and accessible of all the sciences. So, I have tried to explain basics and concepts in simple English where appropriate and without a glossary. But, *caveat emptor*, as Roman engineering geologists might have written in correspondence from Hadrian's Wall to their legionary superiors in York 2000 years ago.

Then, I thought, would that be enough for those such as my own father? He was a thinking man of great curiosity who frequently exclaimed his wonder, but also puzzlement, at the complexity and diversity of the natural world around him. I thought of our family watching the English comedian, Tony Hancock, in his popular television shows of the early 1960s. At the beginning of one episode, we see him deeply immersed in a concentrated study of a thick book, grunting and grimacing as he does so. Finally, he flings the volume across the room with a cry of exasperated rage and incomprehension: 'It's not *me*; its *him*!'. While not wishing to compare my offering with Bertrand Russell's *The History of Western Philosophy*, I hope it won't trigger such a reaction in my readers.

So, my aim for the book is to tell of the discoveries made by my choice of thirteen pioneers of earth sciences. I do this by concentrating on biographical details (where available) and of direct quotes from the original science presented in their papers and books, without too much editorial comment. The book's coda offers more as it tries to bring selected general topics forward to more recent developments. As a devoted reader myself, it seems that one usually learns more from original stuff than from attempts at paraphrase.

The chosen pioneers were all born in the latter half of the nineteenth century – their discoveries coming thick and fast as the new century unrolled. Most of them were young when the awful catastrophes of the twentieth century began to roll out – the First World War and its aftermath of revenge and revolution. Its even worse successor enveloped many in both passive and active service to their diverse countries and in their personal lives. One can only marvel at what they achieved, having lived through such terrible times.

By 'pioneer' I mean 'prime originator', or 'mover'– extraordinary individuals who applied their diverse scientific expertise to increase the understanding of the structure, composition and dynamics of the earth – in making original and lasting theories both by direct observations of its superficial parts and by remote sensing its deeper ones. Some of them, notably Beno Gutenberg, Alfred Wegener and Arthur Holmes, heralded and forwarded the concept of *Our Mobile Earth* (1926) – the title of an influential book by that thoughtful Canadian geologist, Reginald Daly (1871–1957). Like Daly, they mostly made their contributions in the first half of the twentieth century. Notable exceptions were Patrick Blackett's work on rock magnetism, which came about from his post-World War II interest in the deep origin of the earth's magnetic field and Ralph Bagnold's long-term interest in the fundamental mechanics of sediment transport by turbulent fluids.

I must stress here that I am not attempting a *general* history of today's earth science by underplaying the efforts of the countless individuals responsible for the essential basic geological fieldwork that proved necessary for the state we are in today – those who know me will be aware of my own devotion in that department in northern Britain, Greece, Tibet and New Mexico. Rather it is about individuals who refreshed an older science by infusions from widely varied fields of scientific enquiry.

Following Newton's lead ('If I have seen further, it is by standing on the shoulders of giants'), the pioneers did not just spring into the limelight *sui generis,* self-formed. Each had forebears whose shoulders they had stood on, but perhaps more importantly in view of the novelty of their discoveries, they encouraged younger climbers who continued improving their pioneering routes and free-climbing ascents. All three categories – pioneers, precursors and followers – will be discussed in this book, though with greatest emphasis on the original work of the pioneers themselves.

Readers interested in the wider history of geology can turn with enthusiasm to Mike Simmons' *Great Geologists* wherein he develops neat cameos of 35 individuals from the seventeenth century all the way to the present day. Also, how another geologist, Graham Park, sees the main developments over more recent times as revealed in his *Breakthroughs in Geology,* a subject-led account of mainly tectonic themes that nicely complements some of my own.

To my great regret, only one female pioneer, Inge Lehmann, is identified – a telling indication of the now well-known extreme gender bias against female participation in civil, artistic and scientific society that continues in many benighted lands. To take relevant British examples, the Geological Society of London took 112 years from its foundation in 1807 to grant Fellowship rights to women. Since 1919 it still has had only a handful of female presidents, with a mere 19 females being awarded its most prestigious medals (Wollaston, Lyell and Murchison) out of a total of 303. As I write this, optimism is encouraged by the fact that there is a serving female president and that more than 40% of all female medallists were recipients of awards made since 2016. Doubtless this has been due both to increasing female access to the science and to the many and varied careers open to its modern practitioners, but also to the increasing gender equality of the Society's governing council – as reflected by the voting wishes of an increasingly female and, perhaps, a more socially responsible attitude by male Fellows.

Regarding the Geological Survey of Great Britain, ex-Director E.B. Bailey writes in his 1952 history that it was as late as 1943 when the first 'lady' (his descriptor) geologist was appointed to that organization. This was Ms Eileen Mary Guppy B.Sc., (future MBE), ex-Bedford College, University of London. Bailey neglected to mention that she was demoted from that position in 1945 in favour of returning males demobbed from the services during post-war austerity.

Forewords

Gradually, as the twentieth century began, the very tenets of traditional geology were enriched and transformed by the most fundamental scientific advances since the Newtonian Revolution some 250 years earlier. New theories also emerged for earth behaviour gained from traditional geological field evidence, notably the evidence for a c.300-million-year-old glaciation over all of the present-day southern continents in Wegener's 1912 concept of the lateral mobility of continental masses over geological time. This was his *Verschiebung*, literally, to 'shift' or 'journey'; in the 1922 English edition translated as 'drift'. Here are a few general examples of the new advancements:

1   The physical principles governing the flow of heat and of energy exchange (the so-called three laws of thermodynamics) established in the nineteenth century had already enabled first efforts in understanding the age of a cooling earth. Now they were applied to chemical reactions and changes in state such as those involved in the melting of rock and in crystallization from the molten state.

2   The recognition and partitioning of vibratory earthquake waves by dedicated and sensitive instruments on land enabled investigation of the structure and subdivision of the vast extent of the hidden planet – its crust, mantle and core. These were established as primary layers of contrasting physical and/or chemical composition.

3   The discovery of emanant rays generated by electrical discharges and spontaneously by unstable radioactive elements allowed the absolute age of certain rocks to be determined. Latterly, radiocarbon dating of organic remains created younger timescales that revolutionized archaeology and post-50 000-year stratigraphy.

4   The discovery of the stable isotopes of lighter elements such as hydrogen and oxygen and their natural fractionation and preservation enabled a revolution in the study of past environmental change – in atmosphere, land and ocean.

5   Identification of rock magnetism and the seeming permanence of a pole-orientated (dipole) magnetic field led to theories for its origins. The design of ultra-sensitive kit enabled the determination of remnant magnetic vectors preserved in rock – polar declination, latitudinal inclination and field reversals. This led to confirmation that continental drift *had* taken place and, with the aid of radiometric dating, to the subsequent development of an independent magnetic reversal time scale.

6   Aviation stimulated investigations into the dynamics of fluid flow, enabling explanations of the shear and, especially, lift forces generated along solid boundaries to flow. These led on to the development of loose-boundary hydraulics and of lucid explanations for sediment transport and sand accumulation as dunes and ripples by wind and water.

7   Nascent orbital theory for climatic variations over geological and archaeological time received a firm numerical shot in the arm with the advent of algorithms to solve fundamental questions concerning the variability of solar radiation (insolation) over tens to hundreds of thousands of years – the first reliable 'summer-sunshine' records that could explain the ice ages.

The full consequences of such developments came to fruition in the 1960s – the result of the application of these and other discoveries in the basic sciences over the previous six decades. There were also many technological innovations made during and after World War II, notably in computing in its use in the rapid and remote exploration of the ocean floor by sonar. Before this only depth soundings by cable were available. Deeper foundations were revealed by interpretations of soundings by gravity and magnetic measurements, also by the global record of earthquake waves recorded on land and, later, in the oceans. The eventual result, when mixed with the heady fruits arising from the confirmation of the concepts of continental drift through palaeomagnetism and sea-floor spreading, was the plate-tectonic revolution by 1968.

The background to the origin of such advances – the efforts of the 'pioneers' who changed geology to earth science – has been largely neglected by historians of science in favour of often-arcane issues in eighteenth- and nineteenth-century geology or with the plate-tectonic revolution itself, which culminated in the late 1960s. It is often forgotten that earth science is more than just plate tectonics, just as physics is more than just atomic theory. The present book concerns the overlap between the two points of view – a liminal view as it were, of the margins between the nineteenth-century legacy of Charles Lyell and today's vision of our planet as a recycling entity involving all three phases of matter – solid, liquid and gas.

Some would argue that whatever name is given to the study of the earth and its history – geology, geological science, geoscience, earth science or environmental science – matters little, and that the topic should be dismissed Shakespearian-style, as in Juliet Capulet's famous dissing: 'What's in a name?' This author's view is that names *are* relevant – traditional geology was overwhelmingly historical while the new kid on the block, earth science, had something extra on offer. This was the practice of what the 3rd Lord Rayleigh memorably referred to as 'outdoor physics' (and, by extension, chemistry), applied directly to the earth – our present-day geophysics and geochemistry.

Since the days of James Hutton in the late eighteenth century, geological logic has always depended on the application of the basic rules of the 'true cause' (*vera causa*) principle that was used by Isaac Newton in his analysis of how gravity controlled the workings of the solar system. He had realized that the gravitational force involved in general motion could only be experienced, measured and calculated by the *observed* reaction and behaviour of stationary, steady or unsteady moving bodies. These could be falling apples, orbiting planets and moving sand grains, and, as doubtless experienced daily by the older Sir Isaac, in the abrupt jolts of his Sedan chair carrying him around in Queen Anne's London. The exact cause of gravitational force and of universal 'dark matter' still puzzles physicists.

Applied to geology, 'true cause' logic enabled James Hutton and Charles Lyell to infer past processes unobserved by humans and whose actions could be recognized only by the geological outcomes of their passing. Examples included the nature and stratification of sedimentary deposits; crystallinity of lava flows; magmatic intrusion at high temperature; stratal faults and folds, and so on. As earth sciences developed, the emphasis shifted to investigating the additional physical and chemical evidence behind such events and, more importantly, of the mechanical and thermal reasons for them.

Earth scientists still assume that presently acting physical and chemical processes are essentially the same (in kind if not always in magnitude) as those active in the past. Indeed, Lyell understood that the quality of explanation for geological events depends ultimately on the degree of understanding of physical and chemical processes acting on and within the earth. For example, in the first edition of his *Principles of Geology* (1830–32) he proposed a fluid dynamics experiment to his friend, the pioneer volcanologist George Poulett Scrope, which might shed light on the process of sediment transport and the development of rippled bedforms. His experimental design, a paddle-driven water channel, was basically the same as that used by Ludwig Prandtl some seventy years later to illustrate new concepts in boundary layer flow. Hutton, by way of contrast, ridiculed his uneasy acolyte James Hall's designs for the experimental melting of rocks – referring to them in a sarcastic way as merely the results of Hall having 'looked into the bottom of a little crucible'.

It is therefore through the lens of fundamental scientific discoveries that we must view earth science – it was evidence from physicists, chemists and engineers that led to its beginnings. A good number of the persons involved would not have considered themselves geologists at all – happy to bear the name of their mother disciplines in physics, fluid dynamics, chemistry, thermodynamics, engineering, rock and soil mechanics, etc.

Environmental science is a mid- to late-twentieth-century discipline that developed from earth science (with the addition of meteorology, climatology, ecology and, in some places, sociology). Its rationale as the study of ambient Earth, familiar and home to *Homo sapiens*, also involves geographical, social and ethical disciplines, all of which are valid approaches in the wider scheme of things. Such human-centred science certainly does not exclude contributions from geophysical studies of the inner earth that have obvious surface repercussions in terms of the major natural hazards inimical to mankind – volcanic eruption, earthquake, landslide and tsunami. The very best environmental science schools have large interdisciplinary faculties that span the whole gamut of ambient earth science processes and their effects on us humans – from roiling outer core to slowly convecting mantle, mobile plates, blowing wind, flowing river water, surface waves and shallow and deep ocean currents.

PART 1

Deep Stuff: Settings

Richard Oldham (1858–1936); Andrija Mohorovičić (1857–1936); Beno Gutenberg (1889–1960); Inge Lehmann (1888–1993)

James Hutton's game-changing *Theory of the Earth* (1788, 1795) involved a grand scheme of repeated heat-induced crustal elevation and depression. Yet little was known for sure at the time regarding the temperature or state of the deep earth. In fact, the Comte de Buffon had already made temperature measurements in deep mines, finding an increase with depth, so there was room here for plenty of speculative and fanciful stuff, some of it inherited from Aristotle – cavities of steam, underground fires, later the idea of vast coal seams stoking volcanoes, and so on. The distinguished chemist and keen geologist, Humphry Davy, even speculated in 1805 that his beloved alkali metal elements, calcium and potassium, might produce such conflagrations when tapped by percolating water.

The first breakthrough came just a year after Hutton's death with Henry Cavendish's 1798 'Experiments to determine the density of the earth' published in the *Transactions of the Royal Society*. He improved a pivoting apparatus of suspended lead balls devised by the Reverend John Michell who had 'contrived a method of determining the density of the earth, by rendering sensible [visible] the attraction of small quantities of matter…'. This exquisitely accurate piece of kit and use of Newton's general expression for the gravitational attraction of bodies enabled Cavendish to establish an accurate estimate of whole earth density – 5480 kg m^{-3}.

Emil Weichert of Göttingen (the world's first entitled professor of geophysics) pointed out 100 years later that Cavendish's result was much denser than any known crustal rock made up of silica-based minerals. It was almost twice that of the granitic continental crust; around 1.9 times that of the basaltic ocean crust, and in the range 1.6 to 1.7 of the mineral olivine that made up a good part of his underlying 'Stein mantel' (rock mantle). He reasoned that the earth's deepest parts must be much denser to arrive at the mean density established by Cavendish, and had the brilliant idea that maybe the earth was just like the products of the blast furnace – raw metal surrounded by fused silicate 'slag'. Even more presciently, that it might consist of concentric shells – a huge metallic core of iron-nickel (densities 7800 & 8900 kg m^{-3} respectively) surrounded by a silicate-dominated rocky mantle. Using Cavendish's result, he proposed a ratio of core to mantle radius of between 3 and 4 – the order of magnitude to be expected for compressed iron in the core with a radius of *c.*5000 km surrounded by *c.*1400 km of rocky mantle.

This was a verifiable scheme for earthquake scientists (seismologists) to test – seismic waves bringing what Weichert described as 'tidings from afar' to their early and increasingly accurate recording instruments known as seismometers. These enabled tiny three-dimensional motions of the earth's surface caused by the earthquake-generated 'shock waves' to be magnified and recorded on paper charts. Richard Oldham's pioneering account and analysis of the Great Assam Earthquake of 1896 and the subsequent use of seismic waves from other earthquakes by himself, Mohorovičić, Gutenberg and Lehmann firmly delimited and improved Weichert's version of two earth shells. Their various discoveries and eventful lives begin our story.

1

Richard Dixon Oldham (1858–1936)
Anglo-Irish geologist and seismologist

Oldham in 1917, age 59. Copyright image x43663, National Portrait Gallery, London. Reproduced by licence.

He documented the 'Great Assam' earthquake of 1897 in unprecedented detail, determining the velocities of the three types of seismic waves shed from it and identifying the fault responsible in the field (1899). During his subsequent long and active retirement he continued his seismological research and wrote vibrant essays on the modern concepts enriching traditional geology and geography. Most notably he presented critical seismic evidence for the existence and approximate depth of the mantle/core boundary (1906), later reassessing his interpretation of seismic wave arrival times and favouring a fluid core (1919).

A big earthquake in faraway Assam

On 12 June 1897 in the Presidency of Assam, NE India, sporadic early monsoonal rains were falling gently onto tea gardens, temples and monumental statues celebrating distinguished members of the British Raj. Around 5 p.m., tens of millions of humans shared a sometimes horrifying experience that many would remember for the rest of

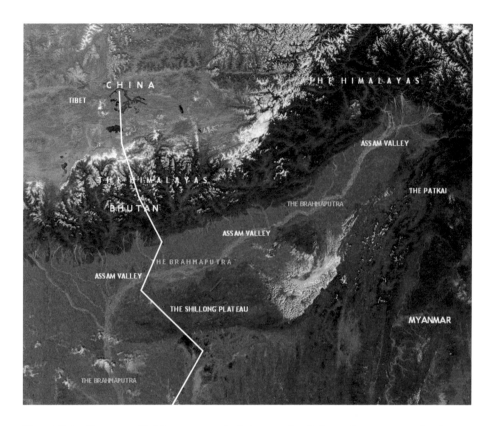

Figure 1.1 Satellite image (E–W extent *c*.800 km) showing the Shillong Plateau as an isolated feature lying south of the Himalaya, with the Brahmaputra River hugging its northern and western margins before its southern spread into the great Ganges–Brahmaputra delta. The white line shows the rough orientation of the northern part of the section in Figure 1.3. The left red line is the rough surface trace of the Chedrang thrust fault that Richard Oldham mapped out in the field.

their lives. Over a large area extending outwards to all adjacent provinces, the surface of the earth suddenly trembled and convulsed as extremely powerful earthquake motions rumbled and rippled across the entire northern half of the subcontinent. They radiated from a central area of most violent agitation in Central Assam along the margins to the Shillong Plateau south of the Himalayan foothills adjacent to the great alluvial valley of the mighty Brahmaputra (Fig. 1.1).

Due to destruction of the regional telegraph system, news of the catastrophic event was slow in reaching both the British Viceroy in Delhi and the organs of the Indian press. Five days later, the *London Times* of June 17th reported the event as Calcutta-based, with due prominence given to the nation's national drink through the damage done to the tea gardens. Next day the paper printed telegrams sent by the Viceroy to the Secretary of State for India in London, reporting details of the damage done in certain districts, indicating that casualties were not high (they would eventually total nearly 2000), but damage to buildings was heavy. Subsequently, more realistic reports appeared that revealed the wholesale destruction of public buildings all over Assam with colossal loss of food supplies and crop damage.

Letters from English-speaking inhabitants and officials living around the Shillong Plateau paint a horrific picture of both devastation and the physicality of a landscape shaken by one of the most powerful earthquakes ever recorded. In Shillong itself there was a *c.*3-minute pre-shock underground rumbling noise before the main long shock, a two and a half minute episode. This caused instantaneous damage in the first 15 or so seconds, according to Mr. F. Smith of Geological Survey of India who was stationed there. All buildings bar plank-built houses collapsed, the Reverend G.M. Davis noting that his church became a heap of stones in less than one minute. Lakes had dried up and huge loose rocks on the ground were thrown upwards in defiance of gravity.

Elsewhere, later than the main shock, on the alluvial lowlands bounding the Brahmaputra River, inhabitants saw consecutive 'earth-waves' passing through paddy fields, the rice stems falling and rising as the waves progressed. Everywhere roads and railway lines were broken, twisted and subsided, with bridges buckling and collapsing. Fields featured innumerable waterspouts rising over a metre high (like 'fountains'), some fissure-like, their effluent laden with loose sand that, visible in the hours following the earthquake, had spread widespread deposits over the land surface, ruining hundreds of thousands of agricultural acres. Organized by the local Commissariat, rescuers toiled that night to search and rescue people amongst strong aftershocks.

Geological Survey of India

It was in the aftermath of this maelstrom of natural destruction that the only professional organization in Imperial India (and perhaps in the world outside of the USA) capable of recording the event was set in action. This was the Geological Survey of India (GSI) whose previous history is worth telling in view of the immense problems set up by the earthquake. It had begun as a 'Coal Committee' in 1836, charged by the East India Company (the *de facto* colonial governing body at the time) with exploring for coal reserves in eastern India. David Hiram Williams, a talented Welsh mapping surveyor from Henry de la Bèche's nascent British Geological Survey, was eventually appointed as 'Surveyor of coal districts and superintendent of coal works, Bengal' in late 1845 on a salary of £800 p.a. Over the next two years Williams mapped immense reserves of coal in the Damoodah-Adjii and Talcheer districts and in February 1848, (the 'Year of Revolutions' in Europe) was appointed Geological Surveyor. Poor Williams was unlucky, for in November of that year while on fieldwork in the Damoodah valley, the GSI records that: 'he fell off his elephant and, soon after, died with his assistant, J.R. Jones, of 'jungle fever' [malaria]'.

Williams' successor was Professor Thomas Oldham, Director of the Geological Survey of Ireland and holder of a chair at Trinity College, Dublin. Late in 1851 he sailed for India with his newly married bride, Louisa Dixon, of Liverpool. Oldham was outside the Oxbridge circle, educated since the age of 16 at Trinity, afterwards studying civil engineering and geology at the University of Edinburgh under the reconstructed Neptunist (all rocks were seawater precipitates), Robert Jamieson. The doughty 35-year-old Anglo-Irishman had vast field experience, high intellect, and was clearly a born leader who at once engaged the East India Company on matters of strategy for the simultaneously established IGS at its headquarters in Calcutta.

The survey needed a convincing mission statement, and its new superintendent provided one in spades. He wanted a British/Irish Geological Survey slant to this important imperial organization – something more than just the narrow aims of a coal-prospecting outfit. He argued that successful exploration for coal should follow, and not lead, the basic scientific geological mapping of the continent – wherewith its natural mineral wealth would fall out in an organized, scientific manner. This would involve field mapping, stratigraphic and fossil studies, setting up a museum reference collection of rock and fossil samples, and the publication of results. This enlightened philosophy governed the progress of the IGS in its early days with Oldham himself contributing both scientific writings and, in his later years, investigating and cataloguing the great earthquakes that had hit the northern subcontinent during historic times.

To the latter end Oldham made a careful study of the large Cachar earthquake of 1869 in eastern Assam. He did so along the lines laid down by fellow Irishman, Robert Mallet, a structural engineer and pioneering earthquake seismologist. Mallet was the first to consider in detail the attributes of earthquakes, author of the influential 1847 'On the dynamics of earthquakes' published in the *Proceedings of the Royal Irish Academy*. He subsequently reported on his observations of the great Naples earthquake of 1857 and introduced the words 'seismology' (the study of the motions engendered by an earthquake), 'isoseismal map' (the spatial range of the intensity of earthquake damage) and 'epicentre' (surface projection of the deep source of an earthquake) into the English language. Oldham brought the records collected by him during his earthquake studies to England on his retirement from the Survey in 1876, but ill health prevented the completion of his report, and the notes were eventually returned to India where, as we shall now see, his youngest son Richard made full use of them.

Richard Oldham's big break

Richard Dixon Oldham, born in 1858, was the geological hero of the entire Assam earthquake investigation. There is little extant information concerning his character and personal life, bar the rather interesting comment of his *Nature* obituarist in 1936, fellow seismologist Charles Davison, which suggests he had inherited more than just his father's drive and initiative; that he:

> was an original and independent thinker – a little too independent sometimes for those in authority...Oldham retired from the Survey in 1903, and for some time lived in the Isle of Wight, where he was near the seismographic station of his great friend, John Milne.

Along with Emil Weichert and Boris Gallitzin, Milne was one of the outstanding instrumental and seismological workers of the late nineteenth century, who in 1895 notably invented what is known as the horizontal pendulum seismograph. This records sideways ground motions, and these seismic recordings were of great use to Richard Oldham in his later studies.

Oldham was educated at Rugby School, close to the family home back in England, then, like so many other geologists, at the foremost geology department in Europe at the time – the Royal School of Mines (later, Imperial College) London. He joined the staff of the Geological Survey of India as a young assistant-superintendent in 1879, and soon afterwards was dispatched for dry-season fieldwork to the Himalayas. One of his earliest tasks during monsoon seasons was the completion of his recently deceased father's memoir on the Cachar earthquake, more than half of which, including the entire discussion of observations, is due to him. He was also responsible for the editing of his father's unpublished catalogue of Indian earthquakes, published in 1882–3 as a memoir of the IGS.

When the Assam earthquake struck on 12 June 1897, Oldham, then Deputy Superintendent of the IGS, was the only geologist at the Calcutta headquarters with a scientific understanding of earthquakes. In his own words:

> The earthquake found the geological survey in a manner unprepared for it. Of the officers available for despatch into the earthquake's shaken tracks, there was but one who had paid any special attention to the subject of earthquakes or had any knowledge of the nature of the observations required, beyond such as might be obtained from the ordinary curriculum of a geological student.

Oldham was referring to himself here, for he was easily familiar with current earthquake theory through his wide reading and previous labours undertaken on his father's writings. He knew exactly what was required in the way of observations after an earthquake event. Four of his colleagues, followed later by a fifth, were quickly transferred from other duties and, after what we must imagine as intense seminars from him to his colleagues, sent to different parts of the central area affected by the earthquake to collect specific field data on damage and other seismic phenomena. A particular innovation was Oldham's decision to seek the help of the public (mostly British expats, but also native Indians) in obtaining relevant information:

Every attempt was also made to obtain information by means of letters and circulars; a large number of volunteer observers were interested in maintaining a record of the aftershocks...

During what he describes as the 'cold weather' of the following winter, six months after the earthquake, he made a field excursion through the whole epicentral area. The long, full and detailed report on the earthquake was published in 1899. It was a great achievement – to bring to fruition so quickly this very first 'modern' post-earthquake investigation that spanned social, landscape, agricultural, architectural, geological and seismological observations with numerous photographs, line drawings, woodcuts and maps. Nothing on this scale had been published before – to read it today is an exercise in erudition – it was the inspiration behind subsequent reports, notably that of the 1906 San Francisco earthquake.

Oldham's report: field observations and human reactions

Memoir 29 of the GSI was published in 1899 'by order of his Excellency the Governor General of India in Council', part of the regular series whose first volume was edited by Thomas Oldham exactly 40 years before. It bears a frontispiece of the twisted and partly collapsed obelisk at Chatak raised in memory of the pioneering hydraulic engineer George Inglis. The field investigations of the IGS team led to information on a plethora of phenomena whose physical explanation required Richard Oldham's close attention. Among the contents are a valuable account of the general principles of seismology; narrative accounts; the determination of isoseismic lines and the area over which the shock was felt; the time of commencement of the shock and the rate and range of motion of the three types of earthquake waves. Landscape features due to the earthquake included a major thrust fault scarp, other earth fissures, sand vents and landslips. Special features investigated included aftershocks, the results of a topographic levelling survey of the epicentral tract (the results were used to re-interpret the tectonics of the earthquake over 100 years later), the position and extent of the seismic focus, records of the Bombay Magnetic Observatory, seismic records in distant Italy of the 'unfelt earthquake', electric effects and earthquake sounds. A particular section investigates the tectonic rotation of man-made structures such as pillars and monuments. Quite a list!

In the epicentral area Oldham concluded that:

> All the accounts...agree in showing that there was a very considerable vertical component in the shock, loose stones lying on the surface of the roads being tossed in the air like peas on a drum, this vertical movement being accompanied by a more marked backward and forward movement of the ground, the sensation produced being generally expressed as being 'shaken like a rat by a terrier'... the range of this motion having been at

least 8 or 9 inches. As to the period, or rate of repetition [it] must have been approximately one second for the double movement forwards and backwards. It may have been a little more rapid, but I do not think it would possibly have exceeded 1½ seconds.

This strong vertical component in the epicentral area was something that Oldham had expected from the nature of the propagating seismic wave pattern. Seen in terms of ground acceleration:

At Shillong, Gauhati, and throughout the epicentral tract, stones have been projected upwards…shows that the acceleration in an upward direction must have been greater than that of gravity… How much greater it is impossible to say but in places it may have been four or five times as great.

As he explains, Oldham's brief in writing his report was that it should concern solely the scientific aspects attending to the earthquake and its aftermath. However, there are several instances in it where the human side to the disaster both for British colonists and the native inhabitants is revealed. I quote here examples from the Garos Hills area at the western end of the epicentral area around the small station of Tura that bordered the alluvial plain to the north. The range of hills was home to the Garos people, whose behaviour during and after the earthquake was less than admirable according to the testimony of the resident Commissioner:

The earthquake caused a regular panic amongst the plains people. Whole villages were deserted for days, while the inhabitants took refuge in the hills. A few Garos took advantage of this to loot their granaries.

The terror induced into the local inhabitants was much increased by their particular beliefs concerning the nature of the world, and indeed of the British Raj. However much we might mock these, they were nevertheless very real for those involved. Oldham wrote:

The Garos generally were thrown into a state of stolid bewilderment by the earthquake. They left their fields, and retired into their village houses to await further catastrophes. The Garos belief is that the world is a square flat body hung up by a string at each corner. There is a squirrel always trying to gnaw these strings, but to prevent it a demon was appointed. This demon, however, neglected his duty, and in order that his attention might not in the future be diverted from his work, he was struck blind. Now that he can't see, the squirrel of course has the best of it, and it is feared that when one or two of the strings are gnawed, the earth will be turned upside down. Another story is that Her Most Gracious Majesty [i.e. Queen Victoria], not content with the last earthquake, has ordered another and more vigorous one to be followed by a cyclone…Had the houses of the European officials in Tura not been wrecked, the Garos would have made up their minds without doubt that the recent catastrophe was the work of the Sahibs, and excited by the wild stories in common circulation they might have given some trouble.

On the alluvial plains to the north of the epicentral uplands of the Shillong Plateau the Brahmaputra and its tributaries, both the natural and the cultivated riverine landscape and outlying villages were overwhelmed by the eruption of sand-filled groundwater through fissures and pipes. Here are two observations made in correspondence to Oldham:

> Many of the villagers' houses…had sunk a good deal, whilst others filled with two or more feet of sand through the doorways…gaps, fissures and the deposit of sands in some of the corn-fields have been so great that it would be impossible for long time to come to bring these fields under cultivation again, or perhaps never at all. In some places an entire village has sunk, and water come upon the surface to the depth of the knee or the loins. The people of such villages had to pass one or two days on crafts made of plantain barks. The water there subsided gradually.

An otherwise practical and factual account sent to the Chief Secretary to the Government of Bengal was from one Babu Hiranmoy Mukerji, of Muktagachha. It gives us another slant, this time a Hindu religious one, of the prevailing feelings amongst the population. Referring to this and previous historic earthquakes, Mukerji writes:

> According to Hindu sastras, such unnatural visitations will frequently appear in this latter end of the Kaliyuga or Iron age. What have already appeared are mere preludes. The real ones are yet in store. The earthquake is commonly calculated by the Hindus according to a formula quoted… 'If you get famine, drought and plague in one and the same year, you get the earthquake that year.' This calculation has indeed been verified.

Modern earthquake investigators the world over have had to face similar demands from local inhabitants for their religious beliefs to be considered.

Oldham's analysis: towards earthquake science

The task of somehow bringing together and summarizing the widespread information gathered concerning the effects of the earthquake and of its mechanical nature became clear in the months afterwards. Oldham knew that earthquake energy would have radiated out from the epicentral area, diminishing as it went. To assess this diminution and to gain a sense of its geographic spread he used Robert Mallet's original ideas concerning isoseisms (isoseists), and those mapped out by C.E. Dutton after the Charleston, South Carolina earthquake of 1886. The latter relied on the relative destructive effects on standing buildings of various constructions in an area relatively densely populated by affluent humans. The parts of northern India so affected were vastly different – immensely larger in area, almost entirely rural, populated overwhelmingly by humans living under the most basic standards of living, the preponderance of flexible wooden plank housing and so on. Oldham had to make do with evidence for earthquake strength gathered from the relatively few urban centres that contained vulnerable stone-built structures and from the verbal evidence of a few stoical, impartial and educated persons concerning sudden overwhelming events whose destructive phases occurred without warning and lasted for just a few minutes. Oldham used the following six criteria for drawing-up his intensity contours:

1. Total destruction of brick and stone buildings.

2. Damage to all masonry or brick buildings.

3. Damage to most brick buildings.

4. Earthquake motions felt by all; disturbance to furniture and loose objects.

5. Earthquake generally noticed, but not severe enough to cause rearrangements.

6. Earthquake noticed by a small proportion of sensitive and well-situated persons.

That he managed to produce a fine map utilizing these isoseist definitions was a major achievement (Fig. 1.2). It showed the distribution of damage over the vastness of the earthquake-felt area of rural northern India.

In the following winter, Oldham made a long field excursion some 200 km north of Calcutta into the Shillong Plateau to make detailed examinations of damage and to survey any field evidence for co-seismic faulting. In the wider landscape he noted the low relief on the ancient metamorphic basement rocks that underlay the high plateau surface and the presence of deep gorges cut both northwards and southwards down the steep east–west plateau margins. At the western end to the plateau, he located and mapped a major NNW-trending fault rupture that he named the Chedrang fault. It could be traced for a minimum length of around 20 km – the ground always pushed up on its eastern side. The ruptured ground and stream bed surfaces on either side of the fault showed large but highly variable displacements – up to 10 m, generally between 3 and 8 m, but sometimes negligible. His photographs of the fresh fault scarp in the Memoir are, I think, the first such images to appear in the scientific literature.

As he reviewed his field observations and the preliminary results of field surveying by levelling that he had commissioned, he was struck by the clear evidence for thrust

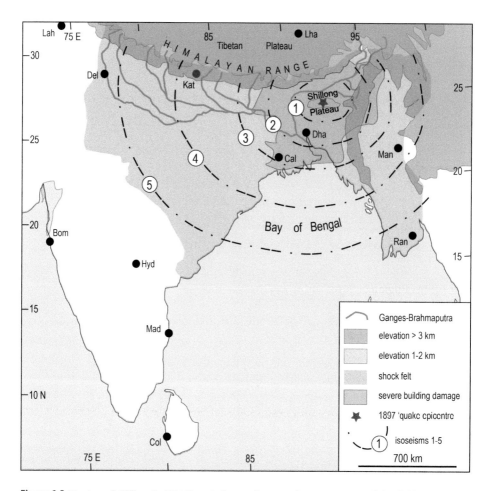

Figure 1.2 Version of Oldham's 1899 'isoseist' map showing the great extent of the 'felt' 1897 Assam earthquake, the spatial variation in damage according to his mapped isoseists and the generalized location of the epicentral Shillong Plateau.

faulting (see Chapter 9 for insights into this phenomenon) and of the long history of uplift of the whole plateau. He made an analogy with the overthrust tectonics demonstrated by the Geological Survey of Scotland along the Precambrian and Lower Palaeozoic rocks of the Moine Thrust in NW Scotland. He reproduced the already classic geological section that passed along the latitude of Quinag in Assynt.

Subsequent authors followed Oldham's ideas on thrust tectonics but assumed the main crustal shortening that resulted was due to an offshoot of the Himalayan frontal thrust system that separates the lowlands of the Indian foreland from the Himalayas and from the Tibetan Plateau far to the north. A radical modern alternative reconstruction of the tectonics behind the earthquake by Roger Bilham

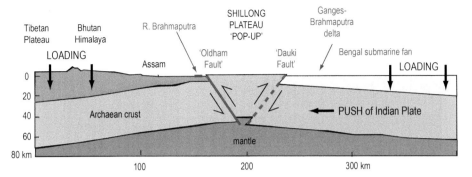

Figure 1.3 Version of Bilham and England's (2001) section to show their interpretation of the Shillong Plateau as a tectonic 'pop-up' structure on two E–W trending high-angle 'blind' thrusts (crypto-faults). No such faults appear to have broken surface in 1897.

and Philip England (2001) was derived from an analysis of much later (and more accurate) geodetic levelling to that undertaken in 1898. It involved the whole plateau as a 'pop-up' structure on a double array of 'blind' thrusts (faults that do not break surface) to the north and south rather than as part of the more general northward underthrusting of the Indian plate (Fig. 1.3). There was no role in this scheme for Oldham's major Chedrang Fault, perhaps a surprising aspect that might deserve further investigation.

'The unfelt earthquake'

There was fundamental science to come in Oldham's report. The huge earthquake had sent its powerful message (Weichert's 'tidings from afar') silently around the globe to the early seismographs of northern and southern Europe, whose records Oldham was able to consult, notably those from the highly respected Italian observatories. The tidings came in three parts:

> The first...was marked by the commencement of the disturbance, the second by a sudden increase accompanied by a change in the period of the waves...I suggested that they represented the arrival of the condensational and distortional waves respectively...while the surface undulations of long period had travelled round the surface of the earth; thus recognizing the presence of the three known types of elastic wave motion.

Oldham's inquisitive mind and widespread seismological reading into the physical aspects of earthquakes led him to examine more exactly the nature of the earthquake that had devastated his adopted homeland. The first (primary) waves to arrive at any distant station, his 'condensational' phase, were body waves whose back and forth 'push-me-pull-you' motions were in the direction of wave travel. These are nowadays known as compression-rarefaction or P-waves. The second, slower (secondary) phase was also made up of body waves that travelled through the earth, but instead of concertina-like motions their passage involved a component of transverse, sideways-shearing motions normal to the direction of wave travel – these are nowadays known as shear or S-waves. As Oldham had previously noted in his Memoir:

> In the case of both these first two phases of the records we have a rate of travel much greater than that of the earthquake which was felt [in India], an increase which must be mainly attributed to their having travelled through the earth at depths where pressure and temperature produced profound modifications in the elasticity of the rocks they travelled through.

More on this below.

Finally, there were the long-lasting, undulatory surface waves that were widely noted during their terrifying 10-minute duration during the earthquake. Oldham ended his account of the 'hidden earthquake' as a conclusion to the main body of his memoir with quietly congratulatory words:

> The recognition of surface undulations which have travelled round the world, may be regarded as one of the most interesting results of this investigation. It is the first occasion, since suitable instruments have been set up, on which the surface waves of an earthquake have been of sufficient size to maintain their character, and leave a recognizable record, after traversing five sixths of the circumference of the globe, and the announcement of this may well bring the report of the great earthquake of 1897 to its close.

Wider insights: velocities of the three seismic wave types

Oldham's tail was up! He had finished his great report and submitted it to the Calcutta printers with the well-earned feeling of a job well done. He had also submitted a paper to the Royal Society in June 1899, the year of publication of the Indian memoir, showing that he was resolutely pursuing a wider goal. In 'On the propagation of earthquake motions to great distances' of 1900 he noted that Mallet and others had all assumed that earthquake motions were due solely to the action of body waves, whereas Lord Rayleigh's 1885 theoretical analysis of elastic deformation had predicted the occurrence of 2D-waves (without divergence) travelling along the free surface of the earth. To further investigate the correctness (or not) of his own deductions he obtained selected accurate records of the Assam arrivals from 11 observatories in Europe (6900–7800 km from India) where advanced seismometer arrays were in place, taking special care that accurate and precise timekeeping had always recorded the advent of the various Assam arrivals.

From these records he was able to confirm in more detail the characteristics previously estimated from Assam. He enquired 1) whether other large distant recent earthquakes (though none were as large as Assam) showed similar patterns of arrivals, 2) whether he could accurately calculate their travel times, and 3) whether he could detect the exact form of the body waves in their passage through the earth to their distant destinations on seismogram paper. Specifically, he was intrigued as to whether the deeply penetrating waves at long distances from the epicentre still showed a linear constant velocity or were accelerated at depth and thus had curved traces in the deeper earth.

He carefully chose seven large-earthquake records distributed up to 120° of arc from their epicentres (the majority around 80–90°). His resulting calculations of travel times, assuming a linear form, showed that P-waves from epicentres at 85° of arc had typical average speeds of nine kilometres a second. S-waves travelled at around 5.3 km/sec and the slower but higher amplitude surface waves at around 3.0 km/sec.

In detail, by plotting arrival times against distance he found a steady increase in seismic velocity for both types of body waves but no such change for surface waves. This meant that the body waves were passing through progressively more elastic earth materials (perhaps denser due to greater pressures), the waves assuming a convex-to-the-centre form rather than linearity. However, he doubted whether such trends would continue to the centre of the earth, but could not be sure because he had no records greater than 120° of arc from the earthquake epicentres and only four records at >91°. In his own words:

> … a sudden change is to be looked for where the wave path leaves the stony shell [lower mantle] to enter the central metallic core which may reasonably be supposed to exist… we may take it that the central metallic core extends to about 0.55 of the radius from the centre, or about the same depth as is reached by the wave paths which emerge at a distance of 90° of arc from the origin. It will be interesting to see if observations which may be obtained hereafter at distances of more than 90° of arc bear this out.

Indeed, and they did.

Major result: discovery of the core

Oldham never wrote about the progress of his research activities following his retirement in 1903 from India due to ill health – a common enough outcome for the men and women of the British Raj and a boom time for retirement real estate in sunny southern England – but he must have long pondered over the nature and size of the core.

Then, in 1904, an ancient, geologically inclined polymath, the Reverend Osborne Fisher, published a thought-provoking paper speculating on the existence of a molten iron core. It may be that Fisher's speculations spurred Oldham on to complete his seismic analysis, writing it up as a landmark paper, *The Constitution of the Interior of the Earth, as Revealed by Earthquakes*, read at a meeting of the Geological Society of London on 21 February 1906, just two months prior to the great San Francisco earthquake. He introduced his contribution with an admonition (perhaps with Fisher in mind) and an inspired analogy:

> The object of this paper is not to introduce another speculation, but to point out that the subject is, at least partly, removed from the realm of speculation into that of knowledge by the instrument of research which the modern seismograph has placed in our hands. Just as the spectroscope opened up a new astronomy by enabling the astronomer to determine some

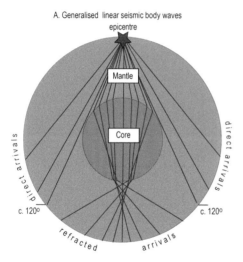

A. Generalised linear seismic body waves

B. The two curved seismic body-wave types

Figure 1.4 Versions of Oldham's 1906 figures illustrating A. Idealized linear seismic ray paths between an earthquake epicentre and a few representative receiver stations located at increasing degrees of latitude out from it. At 120°–130° of arc, extending to the antipodes, all rays recorded by receiver stations are diffractions through an abrupt mantle–core boundary. B. More realistic ray paths with curved trajectories split between primary and secondary P- and S-waves. Note the passage of postulated S-waves through the core (Oldham later saw these as surface-reflected S-waves – see text and caption to Figure 1.5).

of the constituents of which distant stars are composed, so the seismograph, recording the unfelt motions of distant earthquakes...as if we could drive a tunnel through it and take samples of the matter passed through.

With the aid of ten more recently reported earthquake records of suitable attributes whose epicentres were located at arc lengths of greater than the 90° around the globe gained from state-of-the-art recording observatories (many using John Milne's seismometers) he was able to demonstrate again the accelerating nature of body waves as they passed through increasing distances from epicentral motion. He presented two versions of his evidence for the passage of body waves through the earth (Fig. 1.4).

He also showed evidence for an increase in velocity of first phase P-waves and a well-marked stepwise change in what he had interpreted as second phase S-waves for distances greater than 120° (Fig. 1.5). This was the evidence he was looking for – the phenomenon of refraction from and around a dense core creating a 'shadow zone' between 120° to 150° – a definite indication that the core both existed (he roughly estimated its radius around 40% of the whole) and that it had an abrupt boundary with the stony earth layer (mantle) above. He was ultra-cautious, however, about Fisher's

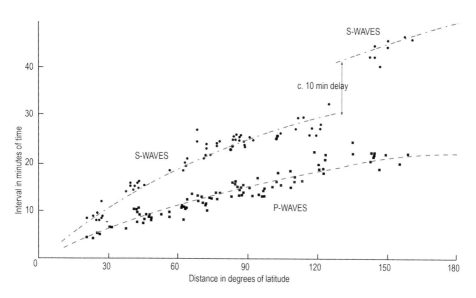

Figure 1.5 Version of Oldham's 1906 plot showing seismic wave arrival times for 13 major earthquake events from a wide latitudinal spread of receiver stations. Note the increased travel time for the more distant P-wave arrivals and the time lag (*c.*10 minutes) for the S-waves after about 120–130°. The latter were later deemed by Oldham (1919) to have been mistakenly identified: they were more likely to be S-wave reflections from the surface. This reconsideration, shared by Gutenberg (see text and Chapter 3) led him to accept a fluid state for the core.

molten core hypothesis, but accepted it as a possibility that deserved, and required, further study. At any rate, just a few months after Einstein had revolutionized physics, Oldham had discovered evidence for the existence of a dense and probably metallic planetary core.

Oldham's retake, a fluid core and the Göttingen School

Twelve years later, in 1918, Oldham had something more to say in general about his discoveries and of their relevance to explanations for the mechanisms invoked by mathematical physicists. In the meantime, studies for a doctoral thesis were published in 1914, the year of the outbreak of World War I, by a brilliant young German geophysicist, Beno Gutenberg (see Chapter 3). He and colleagues at Weichert's geophysics school in Göttingen had concluded that Oldham's interpretation of the seismic wave arrivals from greater distances beyond 120° from the epicentre was problematic. Here is Gutenberg writing 45 years later in his historic summary of the early seismic revolution instigated by Oldham and continued vigorously at Göttingen:

> Oldham had realized that his 'transverse [S] waves through the core' of 1906 are actually transverse waves reflected at the earth's surface (SS) which have not entered the core...

It appears that Oldham was unaware of Gutenberg's pre-war work, possibly due to an understandable lack of wartime scientific communication across the Channel. He reassessed the 'core problem' in a keynote introductory address at a symposium arranged by the British Association for the Advancement of Science held in the rooms of the Royal Astronomical Society on 19 November 1918, just eight days after hostilities with Germany had ended. The address was subsequently published in the *Geological Magazine* for 1919. In the first paragraph of the published address he made some lukewarm comments about the theoretical work of the Göttingen school, going on to memorably illustrate the strongly observational content to a subject that he had himself helped create twenty years before in India:

> When I received the invitation to open this discussion my first feeling was one of diffidence, for, the interior of the earth being necessarily inaccessible to direct observation, the solution of the problems connected with it has principally been left to mathematical research, and this must be the final court of appeal. In these circumstances it seemed verging on presumptuousness to address an audience consisting so largely of mathematicians in inauguration of a discussion on the interior of the earth. Second thoughts showed that there was much to be said against this view, for, though mathematics is the court of appeal, it can only decide on the facts placed

before it by the sciences of observation, and so the discussion seems profitably prefaced by a statement of the leading facts which have been collected, and those conclusions which are so directly derived from them as to have almost the validity of observation.

He went on to review the observations made in his 1906 paper, stressing first the steady increase in wave propagation of both wave phases emerging up to 120° from their epicentres that must have passed through fairly homogenous material of an 'outer shell' (mantle) subject to a gradual change of elasticity due a steady increase of pressure and compression. Arriving at the problems associated with the >140° seismic arrivals he writes:

Beyond this depth there is a rapid transition to a material which can only transmit the condensational [P] waves at a somewhat reduced rate, and is either *incapable* of transmitting the distortional [S] waves, or transmits them with a reduction to about half the velocity attained in the lower parts of the outer shell; at that time [1906] it was impossible to decide between the two alternatives which were both presented, *with some leaning towards the former.* [this author's italics]

After discussing some early theoretical results by Weichert and colleagues at Göttingen on their 'discovery' of a much shallower discontinuity within in the mantle (around 1500 km deep) Oldham latched on to Weichert's 1897 argument for a metallic iron core, accepting his reasoning from density arguments as entirely 'possible'. Regarding the puzzling arrivals after 120°, he repeated the uncertainty he felt in 1906 concerning their true nature, pointedly referring to later Göttingen work which assumed these were in fact true S-waves. He then drew attention to a 1913 paper by G.W. Walker, who had demonstrated that these errant wave arrival times were consistent with their origins as reflected S-waves: they were 'reflected at, or near, the surface of the earth…and this seems still the most probable interpretation.' Noting also recent work done by H.H. Turner and by his own re-examination of seismic records from Italy and the new Gallitzin seismometer at Eskdalemuir, Scotland, he concluded that the irregular form of the >140° arrivals was utterly distinct from the nature of the true S-waves from the mantle shell. These had distinct arrivals in a well-defined packet containing a definite maximum and a more gradual diminution – a record of a single group of waves of one character and form. By way of contrast, the >140° arrivals had a record that:

…bears the impress of being due to the successive arrival of more than one group of waves, just the appearance, in fact, which would be anticipated from Dr Walker's interpretation.

Oldham concluded:

> ...that the wave paths which penetrate deeper than the outer limits of a central nucleus [core], extending to something less than half the radius of the earth from the centre, encounter a material so devoid of rigidity even against stresses of only a few seconds duration.

And summed up thus:

> ...we have found that the interior of the earth is divided into three distinct regions, characterized by differences in the physical condition of the material. They are:
>
> 1. An outer crust, of matter which is solid in every sense usually attached to the word...
> 2. A shell [mantle]...consisting of matter which neither the term solid nor fluid can be applied without introducing a connotation which is contradictory to some of its properties, for while highly rigid as against stresses of short duration, or even of a duration measured by years, it is capable of indefinite yielding to stresses of small amount if of secular duration. At its lower limit this passes somewhat rapidly, but more gradually than at the outer limit, into
> 3. A central nucleus [core] consisting of matter having little or no rigidity, even against stresses of very short duration. Here the material may be described as fluid...
>
> In composition, as distinct from constitution, the earth appears to consist of two parts; a central portion mainly metallic and principally iron, extending to somewhere between three-quarters and four-fifths of the radius, and an outer envelope composed of stony [mineral silicate] material.

He ended on a valedictory and challenging note concerning, once again, the necessity of using surrogate observational evidence (from seismic waveform analysis) for our knowledge of something we cannot directly observe:

> Such, briefly, are the conclusions which may be drawn from the sciences of terrestrial observation. The statement, I know, is incomplete and imperfect; some at least of the conclusions will doubtless be traversed and regarded as incompatible with the results from other lines of research, but in their main features of the threefold division of physical conditions and the twofold division of chemical composition they seem to me to be so well founded that the burden of proof lies with those who would traverse, rather than with those who are prepared to accept them.

2

Andrija Mohorovičić (1857–1936)
Pioneering Croatian seismologist and meteorologist

Mohorovičić in 1910, age 53. Image from Creative Commons via Dr Marijan Herak, University of Zagreb.

He used local and Europe-wide seismic records of a shallow-depth Croatian earthquake with its epicentre close to the capital Zagreb to infer that body waves passing through the regional continental crust were both reflected from and refracted along an underlying rock layer of much higher density. The strong compositional interface became known as the Moho in his honour – an abrupt discontinuity separating low-density, silica-rich continental crust above from high-density, silica-poor mantle below (1910).

Introduction: crustal matters

By the closing years of the nineteenth century and into the early twentieth, enough field and marine geology had been done worldwide to establish that the rocks comprising the continents and those underlying the oceans, were of substantially different composition. The continents comprised generally silica-rich, relatively low-density (*c.*2750 kg per cubic metre) rocks of the granitic clan. By way of contrast, the sub-oceanic crust comprised predominantly silica-poor and higher density rock (*c.*2900 kg per cubic metre) of the basaltic clan. These compositional and density contrasts later gave rise to Eduard Suess' widely adopted mnemonic terms 'sial' and 'sima'. Both rocks contain common silicon, the sial also aluminium-rich, chiefly of silicate minerals like quartz and feldspar; the sima was silicon-poorer and magnesium-rich, chiefly comprising feldspar and pyroxene, with more olivine in some.

By this time, igneous rocks were universally attributed to the products of crystallization from melts of two kinds. Either they were from high-viscosity and deeply buried, slow-cooling magma (mostly granitic) or to rapidly cooling and lower viscosity surface lava or near-surface equivalents (dykes and sills), mostly basaltic. The question arose as to why this marked contrast should exist at all. It seemed that either two processes of melting must be involved and/or there were two contrasting source rocks for melting occurring under the continents and oceans – geologists were used to this kind of messiness in their subject. Several canny individuals realized that it was possible that the less dense sialic continents might actually rest on top of the denser sima – floating like icebergs and following the principles of buoyancy (see Chapter 3). But this implied that the continents were formed from and therefore younger than the oceans! And the analogy with icebergs implied a very sharp compositional boundary between sial and sima.

Oldham's pioneering studies of earthquake seismology raised hopes in the early twentieth century that further progress might see the discovery of more internal partitions to the earth's outer rocky shell. A negative legacy from Richard Oldham's original work (not his fault) was that although the seismic signature of giant earthquakes could delimit deep earth structure, they could shed little light on shallower regional or local structures. Oldham was aware of this, and rather ignored any possibility of finding definite shells in the shallower earth, whilst at the same time envisaging a crustal layer 'perhaps a score of miles thick' below which the earth materials were relatively homogenous down to great depths. The deeper and denser materials became known as Emil Weichert's 'Stein mantel'; the term eventually condensed into English as 'mantle'.

What was needed was a detailed seismological study of a moderate magnitude earthquake whose seismic body waves could be clearly recorded by modern seismometers relatively close by, perhaps within several hundred rather than thousands of kilometres away. We turn to the remarkable analysis obtained by that individual, one Andrija Mohorovičić, after the occurrence in 1909 of a severe earthquake close to Zagreb, then, as now, capital of Croatia.

Enter Andrija Mohorovičić

Mohorovičić was a seismologist in the right place at the right time – possessed of the kind of inquisitive, organized, incisive and mathematically inclined brain capable of sorting out the complex kinds of situations presented by seismology in the shallow earth above *c*.100 km depth. His scholarly background was interestingly diverse, a polymath in the best sense of the word – his biographical details revealing a person who had learnt, taught and researched at a variety of different levels during his long life. Not only that, but he was also a public-spirited person who got things done on state and university committees – seismological institutes were set up and financially supported; new laboratories were provided with up-to-date kit; post-seismic field surveys were organized that involved the local populace in assessing earthquake damage.

He was born in Volosko, near Opatija, Croatia in 1857, just a year older than Richard Oldham, the two men even dying in the same year, 1936. In pre-World War I days, Croatia was part and parcel of another imperial enterprise, the long-lived Austro-Hungarian Empire. His father, also Andrija, came from Istria, his trade a specialist blacksmith making anchors. His mother, Marijanee Poščić, was born in Opatija but died soon after Andrija's birth. During his schooling he proved to be a more than competent linguist, speaking English, French and Italian fluently, later learning German, Czech, Latin and Ancient Greek. Later in life these linguistic talents enabled him to keep abreast of Europe-wide scientific advances in geophysics. His interests also extended to higher mathematics, mathematical physics and the natural sciences – he had studied these at the Charles-Ferdinand University in Prague between 1875 and 1878, where his professors included the physicist Ernst Mach, pioneer of ballistic shockwaves (the Mach Number) – an eccentrically minded physicist who believed only in the reality of phenomena and sensations and had no truck with atoms in any shape or form.

On graduation, Mohorovičić taught in high school and then, in 1882, at the Nautical School in Bakar, near Rijeka for nine years, teaching mathematics and physics and developing a deep interest in both theoretical and practical meteorology, setting up a meteorological station and investigating and publishing accounts of the physical nature of cloud movement. In 1891 he transferred to the capital, Zagreb, where he became director of the Meteorological Observatory and continued his research in meteorology, getting his Ph.D. at the university there in 1893. He remains well known in meteorological circles for his discovery of atmospheric roller vortices in the famous Bora north-easterlies that affect much of the northern Adriatic in winter, particularly Croatia.

It was in the years after this that his attention turned to seismology. Although we have no account of exactly why and when he switched his interests from the fluid dynamics of the atmosphere to the solid mechanics of earthquake seismic waves and the structure of the outer earth, his modern Croatian biographers, D. and M. Herak (2010), make the following reasonable surmise:

> Earthquakes had been of interest in Zagreb for some time, as seismicity around the capital was at its long-time maximum ever since the large earthquake of 1880. Indeed it seems that this intense earthquake activity

played an important role in Mohorovičić's decision to shift his scientific interest from meteorology towards seismology around the turn of the century.

In 1893, building on an earlier small nucleus of geologists interested in seismology and the establishment of a national committee assigned to collecting data on earthquake history, Mohorovičić put into motion the business of drawing up a catalogue of such activity. He established a seismological section within the existing Meteorological Observatory that steadily gained a high reputation worldwide, especially from 1908 onwards, with well-timed records from an existing chronometer and a spanking new Weichert seismograph.

Seismological discovery of a fundamental planetary boundary

Croatia lies within a wide zone of active seismicity in the north-western Balkans – related to its rather complex geological history. The area was part of the maelstrom of tectonic activity involved in the ongoing closure of the ancient Tethys Ocean (the Mesozoic forerunner of the Mediterranean), with its many surviving 'microplates'. In Mohorovičić's time there was no general theory available to explain either the location of individual earthquakes or their regional distribution. So, as part of his research programme at the Seismological Institute, he placed three recording seismometers in stations in the Croatian countryside within areas traditionally affected by seismicity. These were able to record the details of the large earthquake that struck in October 1909 in the Kupa valley, near the town of Pokupsko, only 40 km or so southeast of Zagreb (Fig. 2.1) and which caused extensive damage, though few casualties. For such a strong earthquake to occur so close to the capital and its million or so inhabitants required a thorough examination of the event and an attempt to understand its significance and future implications. This is what Mohorovičić set out to do in the aftermath of the Pokupsko event.

During the earthquake, brick and stone masonry buildings, the latter common in the rocky limestone terrains of Croatia, were extensively damaged over a wide area of countryside, but, as in Assam in 1897, oak plank-framed houses were unaffected. Mohorovičić and his staff were well prepared for this scenario, for since 1901 they had sent out questionnaires on postcards to the inhabitants of areas affected by smaller earthquakes with exact instructions as to how to inform the Observatory of the extent of local damage after a particular earthquake. So, immediately after the 1909 event, hundreds of postcards (430) were distributed in a large-scale follow-up operation to assess the regional distribution of earthquake intensity. Such observations of damage on the ground were then collated, with isoseisms defined and drawn up to a value of VIII (severe) on the new scale of observational earthquake intensity developed by Giuseppe Mercalli in 1902, and which now bears his name. By comparison, the 1897 Assam event was X1–XII (extreme).

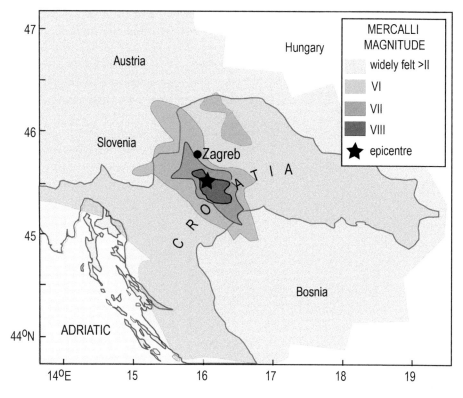

Figure 2.1 A modern reconstruction of the damage zones (according to Mercalli's scale) engendered by the 1909 earthquake south of Zagreb based on data related by affected inhabitants in the days following the event. Simplified after Herak and Herak (2010).

The mapping of the isoseisms located the epicentre of the earthquake. Its location and the accurately recorded arrival time of the initial body wave disturbances at his three recording stations enabled Mohorovičić to make firm progress concerning the interpretation of the seismogram records. In addition, he wrote to the directors of 41 European seismic observatories (from Scotland to Ukraine) for wave records (he made use of 36) and any comments they might have that might help him in his analysis of the earthquake. Amongst the dozen or so records of these correspondents preserved today at the University of Zagreb, various luminaries appear – John Milne, Ludwig Geiger (of the particle counter and a co-author with Rutherford) and Victor Conrad (of the enigmatic 'Conrad discontinuity' within the crust).

Mohorovičić drew up the same sorts of seismic travel-time curves for P and S waves versus epicentral distance that Richard Oldham had constructed. Here was a complete surprise. Both compressional P-wave first arrivals and transverse S-wave second arrivals had a double appearance on certain of the seismic records. Mohorovičić wrote:

> ...the beginning of a travel time of the earthquake – *undae primae* [primary wave] (P) – cannot be expressed by only one curve, there are two curves: one beginning in the epicentre reaching the distances up to 700 km, certainly not beyond 800 km. Second, a lower curve begins certainly at 400 km, but it is possible that it has already started at 300 km... On the basis of the data collected from our earthquake this curve can be drawn up to 1800 km if necessary...

He went on:

> ...when I was certain ...that there are two types of individual primary waves that both reach all places at distances from 300 to 700 km, and... only the first type of waves reach the distances from epicentre to 300 km, while only the second type reach the distances from 700 km, I wanted to investigate this, up to now, unknown fact...There were now two problems to deal with: 1) the estimation of the depth of the earthquake's focus (hypocentre) and 2) explaining the observed pattern of repeated arrivals at distant recording stations.

Knowing the surface location of the earthquake epicentre from the mapped isoseismic lines he attempted to compute the depth of the earthquake hypocentre (focus) – the site of rock rupture along a fault plane. He needed to know this depth in order to interpret the passage of the seismic body waves on their way to his seismic stations. His initial attempts to calculate the hypocentral depth failed for reasons that Richard Oldham had explored in the much deeper-penetrating body waves of the Assam earthquake – the fact that seismic waves curve with depth as they speed up in denser and/or more elastic materials (a mathematical condition known since Newton's time). It would have been no surprise to Mohorovičić that this would also apply to those body waves that traversed the upper crust as well as the underlying 'stony' materials (the mantle) of concern to Oldham. So, giving up the assumption of linear propagation with constant velocity, he proposed the simplest non-linear alternative. Mathematically this needed to gradually increase wave velocity (c) with depth as smooth curves – an exponential function suited the job well – his expression becoming known as Mohorovičić's Law. Theoretical travel-times derived with suitable constants matched the recorded data closely.

It was now necessary to answer the question as to the reasons for the peculiar features of the seismic records (Fig. 2.2). Why did some stations record the arrival of two longitudinal and two transversal waves? Why did such double records not reach stations at distances greater than 700 km (Strasbourg – at an epicentral distance of 720 km – was the furthest station that recorded these). Concerning the former, Mohorovičić wrote:

> ... it is entirely impossible that two different kinds of longitudinal waves with different velocities leave the earthquake focus... both kinds of wave

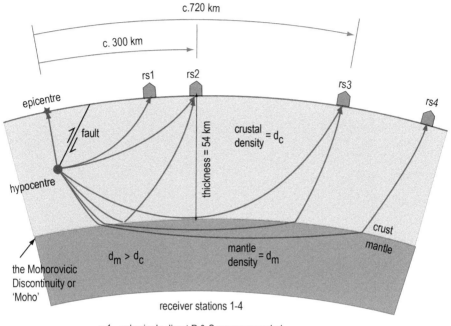

Figure 2.2 Discovery of the shallowest universal internal discontinuity of the planet – Mohorovičić's solution to the puzzle of the 'double arrivals' of direct and reflected waves (at stations such as rs2) and refracted arrivals (rs3,4) encountered within 700 or so kilometres of the epicentre to the 1909 earthquake. After Herak (2005).

are of the same type, that differ one from another only because they reach the surface of the earth by different paths...

He had concluded that, under the outer, relatively thin, rocky shell (crust) to the earth there existed a distinct material whose physical properties differed from the crust in their density, strength and elastic properties. It followed from the laws of physical acoustics that faster travel times beginning along a sharpish boundary would be enough to cause both seismic wave refraction (bending, as in light penetrating water) and reflection (rebound, as in echoes of sound waves). Such an effect would not be possible if the crust continued to great depths as a homogenous medium. A smooth, exponential increase of velocity with depth characterized both media, with wave velocities rapidly increasing on the sharp boundary surface. P and S waves whose rays lie only in the crust were distinct from those entering the mantle and then reflected towards the surface, hence the tell-tale double records.

Mohorovičić then determined the thickness of the upper medium (the crust) above the boundary surface by estimating the position of the rays tangential to the lower, denser medium (the mantle) that reached the furthest horizontal distance (around 720 km) from the epicentre. These rays had travelled at a speed of around 5.7 km/s, typical values for shallower P waves as determined by Oldham. Numerical experimentation showed that the observed data matched the theoretical travel time equations best for a crustal thickness of 54 km. By way of contrast, rays that had travelled in the mantle just below the discontinuity surface by refraction travelled significantly faster, at around 7.8 km/s. Actual reflections of seismic waves from off the boundary surface had theoretical travel-times that matched well with observed data.

Aftermath

Andrija Mohorovičić had the distinction of not only discovering the crust–mantle boundary (later confirmed over much of the earth and named as the Moho in his honour) but also of being the first person to recognize shallow-reflected P- and S-waves. Such reflections, when generated in the shallower sedimentary stratified upper crust by artificial explosions, later became the chief prospecting method in the search for oil and natural gas trapped in various subsurface stratal structures. The Moho itself marked a definite compositional and physical boundary, one that required the mantle to be of markedly greater density and of distinct chemical and mineralogical make-up from the overlying crust. As discussed in the coda to this book, the much later discovery of a shallow Moho under the ocean crust made it clear that the rock that made up the mantle must be of much greater density than even basaltic sima. The only known type that fitted these requirements was ultrabasic (silica-poor) peridotite, a dark, green-black, lustrous rock recovered as mantle fragments brought up in certain volcanoes, mostly comprising the mineral olivine (with associated pyroxene) — the fertile 'mother-rock' of the upper mantle.

Centenary additions

The centenary of Mohorovičić's discovery seemed an appropriate time for two Croatian seismologists, Davorka and Marijan Herak of the University of Zagreb, to celebrate the occasion by revisiting their distinguished countryman's seismic data base. They reviewed the background to the 1909 earthquake using the results of modern geological mapping and of seismological techniques unknown in Mohorovičić's time (Fig. 2.3). In what follows I briefly summarize the conclusions from their 2010 paper 'The Kupa Valley (Croatia) earthquake of 8 October 1909 – 100 years later', published in *Seismological Research Letters*. Some of the tectonic and structural terms used below are explored in the Chapter 9 account of Ernest Anderson's pioneering explanations for the different kinds of faulting.

First up was an appreciation of the regional seismic and tectonic setting of

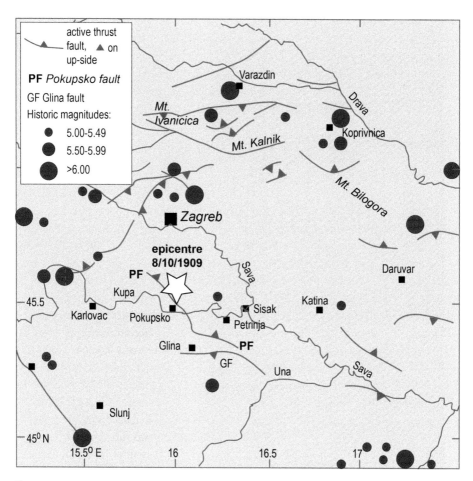

Figure 2.3 The 'centenary take' by Herak and Herak (2010) on the regional tectonic setting of the northern Croatian thrust-and-fold belt, with the location of the 1909 earthquake epicentre adjacent to the Pokupsko thrust fault south of Zagreb. The historic earthquake magnitudes are taken from the Croatian Seismology Catalogue.

the Croatian landmass, located as it is to the east of the Adriatic Sea and of the highly seismic Italian peninsula. The Adriatic itself is relatively aseismic (i.e., few earthquakes, all small) – a stable, undeforming piece of rigid continental plate (Adria) sandwiched between a thrust-and-fold belt (strata folded above low-dipping thrust faults) that is the modern Apennines, a linear mountain chain forming the 'spine' of Italy and a similar thrust-and-fold array in northern Croatia. The Apennines' geologically recent history of compressional (shortening) deformation that formed the fold belt is interrupted today by a new and deadly-active tectonic regime of active crustal extension (stretching) along arrays of normal faults. Northern Croatia is still dominated by crustal shortening and thrusting under compression.

The epicentre of the 1909 event located by Mohorovičić based on earthquake damage was relocated just a few kilometres north, adjacent to a prominent thrust fault mapped in the field and named the Pokupsko fault. It extends into the subsurface as a projection of the 1909 fault's slip plane located at an estimated focal depth of 14±7 kilometres. The magnitude of the event expressed by examination of the national and international seismogram records of the time was estimated as a 'surface-wave magnitude', M_s, of around 5.8. This M_s magnitude scale was first devised by Beno Gutenberg (Chapter 3) in 1942, and for earthquakes of this magnitude is roughly equal to those computed by his friend, Charles Richter, using his more popularly known 'Richter Scale'.

A re-examination and re-assessment of Mohorovičić's post-earthquake damage data, based on the 400-odd postcards from across the region, revealed a 200 km maximum spread along elongated isoseismal lines that run parallel to the structural trends of the major thrust faults south of Zagreb. The sense of movement across the Pokupsko fault itself was obtained by a technique that has become known since the 1960s as a 'fault-plane solution'. This is how modern seismologists remotely determine the nature of the displacement across an earthquake fault – by the sense of its initial ground motion – tension, shear or compression determining the polarity of the very first P-wave arrivals on seismogram records. The Heraks chose archived records of the 1909 earthquake from 11 select observatories to do this. Fault displacement seems to have been an upthrust motion driven by tectonic compression directed SSW–NNE (with a bit of shear added), in agreement with the orientation mapped out for the Pokupsko fault.

A satisfactory modern take to an extraordinary 100-year old discovery!

3

Beno Gutenberg (1889–1960)
German-USA seismologist

Gutenberg in a relaxed pose at Caltech, probably in the mid-1930s. Unknown source. Image provided courtesy of Caltech Library and Special Collections Archive.

After accurately defining the depth (3900 km) to the core–mantle boundary in 1912, he inferred the existence of a subtle but key upper mantle discontinuity, a low seismic velocity zone (Gutenberg Discontinuity or G-Zone) in 1926. He supported Wegener's continental drift theory, amplifying evidence that the discontinuity underlay a rigid lithosphere and overlay a convecting asthenosphere. With Frank Richter he played a key role in expressing earthquake magnitude (1930s–40s) and determined annual world energy production from earthquakes (1956, 1959).

Workings of the earth's mantle – surface evidence

The first clues as to the hidden workings of the earth's mantle came from observations made of surface phenomena. Like so many discoveries in the natural sciences, the magnificence of the world's youngest mountain chains had evoked not only feelings of awe and wonder ('I will lift mine eyes up unto the hills', etc.) but also a great curiosity in the minds of geologically-minded observers, particularly those with a mathematical physics bent. Thus it was that George Everest in the 1840s pondered the implications of the Himalayas rising above the Indo-Gangetic plains for his 'Great Trigonometrical Survey' of India. An anomaly in surveying had been caused by the mountain range's attraction to his surveyors' plumb lines – a phenomenon recognized first by Pierre Bouguer in the Andes, and in 1772 by precise measurement at the mountain of Schiehallion in Scotland by Nevil Maskelyne, Astronomer Royal of England.

Everest's successor as Surveyor-General of India turned for an explanation to John Pratt, Archdeacon of Calcutta and a fine amateur mathematician. Pratt calculated the plumb line deviations that would be expected by the estimated mass of the Himalaya – but the observed deviations were much less than his estimate. To account for this he proposed that the density of the underlying crust was less than that of the surrounding rocks. In a later contribution, he envisaged the range 'floating' on a less solid substrate.

Pratt's proposal was closely studied by George Airy in London, a successor to Maskelyne as Astronomer Royal. Airy also attributed the pendulum effect to a subsurface cause, but envisaged that mountain ranges like the Alps and Himalayas also had a subsurface extent, like piles of logs floating in a river, as he put it. The later analogy of icebergs floating in seawater is perhaps more vivid and immediate.

The Pratt and Airy theories to explain the elevation of mountain ranges both involved subsurface material of contrasting physical properties – in the words of Beno Gutenberg in an end-of-life review written in 1959:

> Airy and Pratt independently concluded that the weight of a rock column of a given cross section above a depth of the order of 100 km is about the same everywhere regardless of the elevation of the surface of the earth. Airy assumed that the greater height of mountains is compensated by greater thickness of the relatively lighter crustal material below them, while Pratt believed that the density of the crustal material under mountains is smaller than that under adjacent lowlands.

Further clues arose from the overwhelming evidence that rapid uplift had followed melting of kilometre-thick ice sheets at the end of the last Ice Age. Charles Lyell, who had at first rejected the idea, documented its effects in southern Sweden when he saw for himself undeniable evidence in the form of uplifted postglacial shoreline deposits around the Baltic coast.

The only viable mechanism for such reverse up-and-down motions was the subsurface flow of a viscous substrate some distance *beneath* the brittle/elastic rocks that defined the Moho. This meant that part of the upper mantle could slowly flow along subsurface pressure gradients, even though it was not itself strictly a proper fluid. A common analogy involved the contrasting behaviour of lumps of pitch, first when hit by a hammer blow and fracturing like any brittle solid, but also slowly deforming by flow if pressed upon long enough by a large enough loading weight.

In 1889 Clarence Dutton, one of a team of brilliant field geologists exploring the western United States under John Wesley Powell (G.K. Gilbert was another), introduced the term 'isostasy' for the situations devised by Pratt and Airy. This word has Greek roots – *ísos* ', equal'; *stásis*, 'standstill'. He wrote in a footnote to his paper that:

> In an unpublished paper I have used the terms *isostatic* and *isostacy* [his spelling] to express that condition of the terrestrial surface which would follow from the flotation of the crust upon a liquid or highly plastic substratum – different portions of the crust being of unequal density.

Joseph Barrell in 1914 envisaged isostasy as that physical state whereby relatively strong crustal and upper mantle 'lithosphere' (Greek, *lithos*, 'rocky') overlies a layer with smaller yield strength, which he called 'asthenosphere' (Greek, *astheneos*, 'without strength'). Below this he assumed that rock strength increased. In his own words:

> The theory of isostasy shows that below the lithosphere there exists in contradistinction a thick earth-shell marked by a capacity to yield readily to long-enduring strains of limited magnitude...To give proper emphasis and avoid the repetition of descriptive clauses it needs a distinctive name. It may be the generating zone of the pyrosphere [where magmatic melts occur]; it may be a sphere of unstable state, but this to a larger extent is hypothesis and the reason for choosing a name rests upon the definite part it seems to play in crustal dynamics. Its comparative weakness is in that connection its distinctive feature. It may then be called the sphere of weakness – the asthenosphere.

These words of Barrell, defining the realms of lithosphere and asthenosphere as independent of the Moho, would echo down the ages to 1967–68 when geophysicists, re-inventing the wheel to some extent, invoked the slippage of lithosphere over asthenosphere in plate tectonics. Meanwhile, it was left to their intellectual ancestors, most notably Beno Gutenberg, to pursue and quantify the nature of the discontinuity that must separate these two rather different realms of state.

Enter Beno Gutenberg

He was born in 1889 in Darmstadt, Germany, to parents whose living involved the manufacture of soap, a product that would figure large in his life in post-World War I years. At the Technical High School in Darmstadt, he developed an interest in the measurement of meteorological phenomena, ignoring his father's wishes that he should become a manufacturer. Learning about new courses in field geophysics provided at the Institute of Geophysics in the University of Göttingen, he enrolled there and developed a deep and abiding interest in all matters geophysical from his

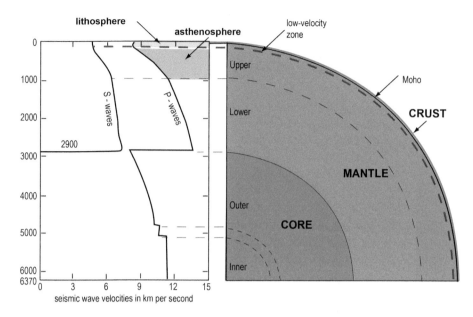

Figure 3.1 Version of Gutenberg's 1959a structure of the earth, the low-velocity zone and variations in velocity of P- and S-waves with depth from the surface with the approximate limits to the lithosphere and asthenosphere.

mentor, the physicist Emil Wiechert. This versatile man had done pioneer work on atomic particles (the electron), his research interests now spanning meteorology, seismology and the nature of the earth's core.

Gutenberg gained his doctorate in physics in 1911 under Weichert (he attended physics lectures by many of the Göttingen luminaries, including Born, Landau and Prandtl) in a study of microseisms associated with ocean waves, a field he would return to 40 years later. His subsequent postdoctoral research at Göttingen established him in his early twenties as a leading figure in world seismology, acknowledged as such by Weichert who told him that he could be taught no more and that he would have to research everything himself.

As we have seen, in 1897 Wiechert had inferred that the earth must have a dense core, probably iron, starting at a depth of about 1400 km while Oldham's 1906 estimate was *c*.3800 km. In 1959, reviewing his important work with Weichert, K. Zoeppritz and L. Geiger at Göttingen in the years 1907–1914, Gutenberg gave an account of German developments in seismology that enabled him to give a much more accurate estimation than either Weichert or Oldham had managed. He pointed out that Oldham's P-waves used to compute travel times were of 'various longitudinal phases', the Göttingen team establishing a firm mathematical basis for the velocity of continuous curved seismic waves, which enabled estimation of more accurate velocities of P- and S-waves in the mantle. Weichert's first re-evaluation of the depth to the outer core using these theoretical functions was much less than Oldham's estimate, *c.*1500 km. Then, as improved seismometers placed in Pacific Samoa began to record deep seismic events that eventually arrived at Göttingen, accurately timed delays indicated that Weichert's estimates from theory were far too shallow. In succeeding years, by studying numerous other accurately determined epicentral events over 80° of arc, Gutenberg reported in 1914 that at a depth of around 2900 km the P-wave velocity changed suddenly from 13.3 to 8.5 km per second, an accurate and precise estimate for the depth to the outer core that has stood the test of time (Fig. 3.1).

Into the maelstrom

In August 1914, along with millions of others, Gutenberg's life was interrupted by war. He served initially in the German infantry but, after receiving a nasty head wound – according to his wife Hertha, death was prevented by his steel helmet – on recovery he became part of a mobile meteorological outfit formed in support of front-line gas warfare operations introduced by Germany at Ypres in April 1915. At war's end he briefly held a position at the University of Strasbourg, losing this when that city became French in 1919 following the Treaty of Versailles.

In the difficult post-war years that followed he had to sustain himself, Hertha and their young family by managing his father's soap factory. The severe and harrowing difficulties of life with hyper-inflation and social tensions in the Weimar republic were movingly recounted 60 years later in an archived interview with Hertha recorded in Pasadena, California (see Gutenberg, H. 1981):

> After the war...there was revolution and great turmoil, and the young people got together and discussed how they could rebuild Germany. Beno was very active in this group, and I was, too. One of my brothers was very active and I had come along with him. That's where I met my husband... We had terrible inflation. That was the worst time, worse than the war. The only [good] thing was that people were not killed; that was the main thing...The mark was valued according to foreign currencies, and in the afternoon it came out—the mark's value relative to the dollar. So in the evening you could get maybe a loaf of bread, whereas if you could get out in the morning, you could get everything for the day....We started to barter. You see, soap was a luxury, and for my boy's first shoes—he was two years old—we bartered with the shoemaker: 'If you give me soap, I'll give you a pair of shoes for your boy.'

In 1926 Gutenberg obtained a junior professorship at the University of Frankfurt-am-Main. It was poorly paid and so he was forced to continue as soap factory manager. Nevertheless, he continued his ground-breaking research in his spare time in his

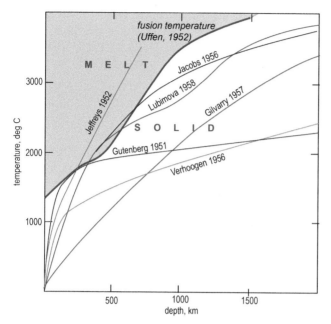

Figure 3.2 Version of Gutenberg's 1959a plot to show temperatures in the upper mantle according to various authors. He shows them in relation to fusion temperatures for mantle materials calculated by Uffen in 1952. Note that Gutenberg's own curve just touches the fusion curve between *c.*1-200 km depth – the source of his low-velocity zone.

home study, establishing and maintaining seismological contacts in the international community far and wide, his work widely cited.

A shallow mantle discontinuity?

One piece of work Gutenberg carried on around this time is highlighted here, neatly summarized by himself, looking back on his research career in 1959 (Fig. 3.2):

> At least as early as 1907, von Wolff had concluded that at depths near 100 km the temperature should be close to the melting point of ultrabasic material [the silica-poor peridotite rocks of the upper mantle below the Moho]. Gutenberg (1926) investigated if earthquake waves show a corresponding appreciable decrease in velocity. He studied the amplitude of longitudinal waves [P waves] of shallow earthquakes as a function of distance up to 3000 km and found that starting at a distance of about 200 km from the source the amplitudes of the longitudinal waves decrease roughly exponentially with distance and reach a minimum near an epicentral distance of about 1700 km, where they increase suddenly... From these results he concluded that there is a slight decrease in the velocity of longitudinal waves at a depth of about 75 km, but no indication of molten material. This was the 'asthenosphere low velocity layer', as he called it later to distinguish it from lithosphere low velocity layers, and has ever since been the subject of investigations with more and more improved methods and results.

Gutenberg may have become interested in this result because of his reading of Alfred Wegener's final 1928 edition of the celebrated *Die Entstehung der Kontinente und Ozeane* (The Origin of Continents and Oceans), first published in 1915 (Chapter 5). Another contact was made with the future discoverer of much deeper seismic boundaries, Inge Lehmann of Copenhagen. She had visited Gutenberg in Darmstadt for a month-long visit in 1927 as part of her self-education in the interpretation of seismogram records (Chapter 4).

By 1930 it seems that Gutenberg had sensed the taint of racial discrimination affecting his many failed job applications to German universities (he was Jewish), and the general political situation made him fear the worst for citizens of his country and his own family. So, before the establishment of the Nazi regime, this world leader in his subject, repeatedly rejected by German academia for posts for which he was uniquely qualified, eventually obtained what can only be described as *the* dream posting. Through international contacts he was offered the founding Director's professorship of the Seismological Laboratory at the California Institute of Technology (Caltech) in Pasadena – it became prestigious under his leadership. He, his family's lives (and American geophysics) were changed for ever.

More seismic evidence

Introducing his final word on the existence of 'his' discontinuity in 1959, Gutenberg gave invaluable background to his own and other contributions to the phenomenon. He introduced both a conundrum and a take on his great near-contemporary, Alfred Wegener (Chapter 5). The conundrum is that 30 years earlier, just before his early death during meteorological fieldwork on the Greenland icecap, Wegener had completed (it was 'in proof' in 1927) the last edition of *The Origins of the Continents and Oceans*. In this final development of his views on the drift (*verschiebung*) of continents, Wegener gave much space to the vexed question of a mechanism by which such rigid bodies could have moved sideways through solid mantle. He cited Gutenberg's 1926 paper and Wolff's analysis with its reasoning that there was both physical and seismic evidence for reduced solidity to the mantle at *c*.100 km depth. He noted this as a possible reason that enabled continental drift to occur. Subsequent investigations by other eminent seismologists (Harold Jeffreys, Inge Lehmann) rather ignored Gutenberg's shallower discontinuity and focused on the recognition of deeper discontinuities of a mineralogical kind (at 200 and 500 km) that would have little direct tectonic influence, but which might determine the thickness of the mobile asthenosphere.

In 1936, Gutenberg began to investigate if recordings of earthquake body waves showed any corresponding appreciable decrease in velocity, as might be expected if 'the von Wolff effect' existed. In 1939, with his friend, Frank Richter, he reported on a study of seismograms from Peruvian earthquakes that utilized deeply-sourced shocks (50–250 km) recorded at short distances (2°–23°) away. Since the advent of plate tectonics, we now know these originate as subduction zone events. The two friends confirmed that the surface 'shadow zone' for P-wave phases previously recognized by Gutenberg was caused by a shallow mantle discontinuity causing a retarded P-wave signal with its greatest extent at a depth of about 80 km (Fig. 3.3). This confirmed and extended Gutenberg's original proposal for the existence of a low-velocity zone, a conclusion further verified by the study of Pasadena earthquakes – both P- and S-wave velocities decreased by 2–3% at depths between 80 and 100 km depth. It seemed clear to Gutenberg that: '…only a relatively small decrease in velocity is required to produce a rather extensive and pronounced shadow zone.'

Subsequently Gutenberg devised a new quantitative method to determine seismic velocities in the asthenosphere using the rate of change in travel time curves. Of 82 shocks originating at depths of between about 50 and 600 km in or near Japan their P and S time curves showed a clear decrease with depth below the Moho, with a minimum at a depth of roughly 100 km for longitudinal waves and 150 km for transverse waves (Fig. 3.3). At the same time the ability of mantle rock to undergo stretching deformation (expressed by a quantity known as Poisson's ratio) decreased. Such results, together with others obtained after World War II using seismogram records of distant Soviet nuclear tests by Frank Press and others, led him to propose a 'channel' model for the focusing of seismic waves along the low-velocity discontinuity (Fig. 3.4).

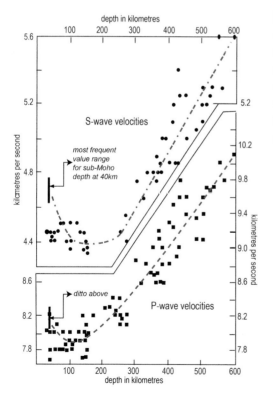

Figure 3.3 Version of Gutenberg's 1959b calculations of P- and S-wave velocities for the upper mantle indicating the existence of a *c.*100 km deep, low velocity zone. As he noted, these indications depend entirely on extrapolations of mean values for velocities at *c.*40km depth in the sub-Moho mantle.

Gutenberg and the route to plate kinematics

As noted above, Gutenberg had long been a convinced mobilist, envisaging initially that centrifugal, non-isostatic and thermal contraction forces might all be candidates for a responsible mechanism to drive continental drift. Later, in the 1950s, he latched onto the idea that mantle convection was probably the main driver, a hypothesis first brought to prominence by Arthur Holmes in 1931 (Chapter 7). At a 1950 conference in Pennsylvania on flow in the earth's interior, the importance of this mechanism was supported by himself, David Griggs, Harry Hess and Felix Vening-Meinesz, each man a leader in his own field. Gutenberg stated:

> …we observe at the surface, phenomena that are possibly connected with convection currents at spots where the currents are going down (or coming up) and that all such observations refer to belts surrounding the Pacific basin…and that the bottom of the Pacific is moving in the same direction relative to the continents in California, Japan, Philippines, and New Zealand.

This was an intuitive view of convection being responsible for mountain building and earthquake activity. Gutenberg's views concerning convection must have been encouraged by the 1950 conference, because convection in the mantle appears as an important component of his thinking in his book *Internal Constitution of the Earth* of 1951, where he suggests that it was probably the most potent of all mechanisms for causing drift. His temperature profile for the mantle (Fig. 3.4) assumed that convection would lower the temperature gradient below the 80 km depth of his beloved low-velocity zone.

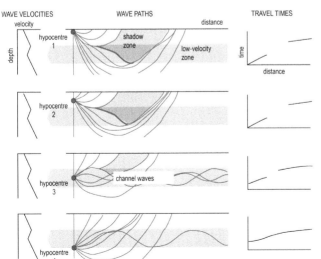

Figure 3.4 Version of Gutenberg's 1959a sketches of body wave paths, travel times, shadow zones and a channel zone for various focal depths if a low-velocity layer exists. The velocity/depth curve to the left is notional. The shadow zone is due to down bend in the low velocity zone of the first refracted rays in scenarios 1 and 2 (heavy red segments).

Gutenberg the man

By all accounts Gutenberg was a generous and kind soul. After the assumption of power by the Nazis he kept his contacts in Germany alive (literally), helping many Jewish scientists to emigrate into the safety of the Anglosphere. Concerning the man himself and his infectious *joie de vivre*, here are the words of his National Academy of Sciences biographer, Leon Knopoff:

> He was small of stature, very personable, and lively. He was well organized and kept to a precise daily schedule. Although his scientific demands on himself were rigorous, Gutenberg was gentle and self-effacing in his relationship with others. He was helpful to anyone who asked a question of him and was tolerant of critics. Gutenberg was a man who could give his colleagues and students a liberal education in scientific method, made pleasantly easy by kindness, patience, amazing industry, and a delightful sense of humour. He was a cultured individual, well read, and with wide interests – reflections of his broad European education.

> Having inherited his mother's musical talents, Beno learned to play the piano, a skill that was to last him through his entire life. In his earliest years in Darmstadt, Beno sang in the synagogue choir and often played the organ. In the Pasadena years, Einstein played the violin in chamber music events organized at the Gutenberg home.

> At the end of World War II, in a reprise of the bartering activity of earlier years, Gutenberg collected the accumulated royalties on his publications in Germany by payment in the form of numerous piano scores for two and four hands. His bookplate shows the Owl of Wisdom with a seismogram in its beak in flight around the Göttingen Institute of Geophysics.

4

Inge Lehmann (1888–1993)
Danish seismologist

Lehmann in 1932, age 44. Creative Commons Licence via the University Library of Copenhagen.

She inferred the existence of a solid inner core inside the molten outer core by using the seismic wave phase, P', picked out from seismograms of a powerful New Zealand earthquake that she interpreted as a reflected wave signal (1936); she later inferred the existence of a 400 km mantle discontinuity, close to the base of the asthenosphere (1960s–70s).

The last frontier

The final breakthrough in determining the fundamental boundaries that divide up Earth's planetary interior came with the surprising recognition of its deepest subdivision – a solid inner core. The tell-tale seismic motions that led to its discovery arrived from the antipodes to be recorded by Inge Lehmann's seismometers in Copenhagen and elsewhere in northern Europe in 1929. After several years of intensive thought and research she eventually published her conclusions in 1936. She had noticed features that no other seismologist had observed, including the inimitable

Beno Gutenberg. According to her biographer, eminent seismologist Bruce Bolt, she was for many years the possessor of perhaps the deepest skills in the far-from-simple interpretation of paper seismic records in those early to middle decades of the twentieth century – before the development of computers.

Lehmann published her discovery in a dark year for the world, 1936, giving it perhaps the shortest title ever seen in the scientific literature: P′. This brisk and economical announcement would have been instantly familiar to the coterie of observational seismologists the world over. In it she writes of her discovery that this waveform arrival, nowadays known in seismological parlance as a reflected PKP wave (transmitted from mantle to core, back to mantle), might have been reflected from the outer boundary to a solid inner core. It had emerged on the seismographic paper of the Gallitzin-Wilip seismometers installed, calibrated and supervised by Lehmann six years previously as part of her supervisory responsibilities for the seismological branch of the Danish geodetic unit, *Gradmaalingen*. In a laboratory space specially constructed within the ancient aseismic walls of Copenhagen – last shaken by Nelson's naval cannon in the notorious bombardment of the city in 1807 – the arrivals of 16 June 1929 provided a breakthrough whose implications thereafter stimulated a huge range of theoretical research. This was in the newly emerging field of the dynamics of convecting magnetic fluids that was to widen the understanding of the generation of the earth's magnetic field.

Who was Inge Lehmann?

She was born in 1888 at Osterbro in Copenhagen, growing up there and living much of her long life on Kastelvej. Her paternal *haute bourgeoisie* family in Denmark had roots in Bohemia – barristers, politicians, bankers, engineers and academics (her father was an early experimental psychologist). Her maternal side featured priests, politicians and prominent early feminists and scientists. She and her sister were brought up in the peaceful atmosphere of the 1890s.

Her secondary education prepared her well for entry into what was then a male-dominated professional society – her parents had enrolled her in her formative years up to the age of 18 in a progressive co-educational school (Faellesskole) run by Hanna Adler, aunt of physicist Niels Bohr and one of the first two Danish women to earn an advanced degree in physics. A prominent educator, she had visited educational establishments in the United States, returning to Denmark to promote co-education. Her civic reputation was such that at the age of 84, Jewish by birth, she was arrested by the Nazis for deportation to the death camps but released after an outcry from the Danish public. Lehmann later reminisced wistfully that she had been encouraged at Adler's school to believe that 'No difference between the intellect of boys and girls was recognized, a fact that brought some disappointments later in life when I had to recognize that this was not the general attitude.'

She writes of her first encounter with earthquakes around the early 1900s when:

…on a Sunday morning, I was sitting at home together with my mother and sister, and the floor began to move under us. The hanging lamp swayed. It was very strange. My father came into the room. 'It was an earthquake,' he said. The centre had evidently been at a considerable distance, for the movement felt slow and not shaky. In spite of a great deal of effort, an accurate epicentre was never found. This was my only experience with an earthquake until I became a seismologist 20 years later.

In 1906 she passed first class in the entrance examination to the University of Copenhagen, which she entered in the autumn of 1907 to study mathematics, her favourite subject, to *candidata magisterii* (master's) level, together with the necessary subsidiaries physics, chemistry and astronomy. In an ambitious venture after passing Part 1 of the examination in the autumn of 1910, she was admitted to Newnham College, Cambridge where she prepared herself for the entrance examinations for the university's Mathematical Tripos. She was forced to abandon this goal after fourteen months due to the strains of overwork and returned to Copenhagen exhausted for a slow recovery. However, she had enjoyed her stay in Cambridge, but writing later she added that this was, 'in spite of the severe restrictions inflicted on the conduct of young girls, restrictions completely foreign to a girl who had moved freely amongst boys and young men at home.'

Her health restored, she used her mathematical skills to work in an actuarial office and in the autumn of 1918 re-entered the University of Copenhagen and graduated in the summer of 1920. The next year she became an assistant to the Professor of Actuarial Science at the University, and for the next three years she became proficient with the theory of observational analysis. In 1925, she was appointed to another assistantship, this time under N.E. Norlund, Director of *Gradmaalingen*, a geodetic institute that had the task of measuring the meridian arc in Denmark. He was planning to have seismological stations installed in Copenhagen and at Ivigtut and Scoresbysund in the Danish dependency of Greenland.

She wrote of her role in Norlund's institute in these early years of her long career in Copenhagen and of stimulating sabbatical visits across Europe:

I heard for the first time that knowledge of the earth's interior composition could be obtained from the observations of the seismographs. I was strongly interested in this and started reading about it…I began to do seismic work and had some extremely interesting years in which I and three young men who had never seen a seismograph before were active installing Wiechert, Gallitzin-Wilip and Milne-Shaw seismographs in Copenhagen and also helping to prepare the Greenland installations. I studied seismology at the same time unaided, but in the summer of 1927, I was sent abroad for three months. I spent one month with Professor Beno Gutenberg in Darmstadt. He gave me a great deal of his time and invaluable help. I paid short visits

Figure 4.1 Lehmann with unknown colleagues on one of her annual visits to a Danish seismological station in Greenland in the early 1930s. Wikipedia Commons via the Carlsberg Foundation, Copenhagen.

> to Professor E. Tams in Hamburg, to Professor E. Rothe in Strasbourg, Dr van Dijk in De Bilt and Dr Somville in Uccle. Everywhere I was very kindly received and given much information.

In this roundabout but personal and committed way, seismology became her career choice. In the summer of 1928, she obtained her degree in seismology and was appointed head of the seismological department of the Gradmaalingen, a post she held until her retirement in 1953. This was not a research post but a practical one, concerned with the upkeep of the three internationally important seismographic observatories in Copenhagen and Greenland, the latter run by caretakers and linked to Denmark and Lehmann's supervision only by an annual boat voyage (Fig. 4.1).

At the Copenhagen observatory she was free to do scientific work, but it was not a duty – her role involved the day-to-day routine of a laboratory superintendent. Entirely under her own steam she fashioned an independent research career from her only academic task – to interpret and publish the Danish observatory seismogram records as a key part of the informal north European seismological network – from whence grew her immense expertise in seismogram interpretation.

At the very beginning of her research she tackled the problem of the accurate location of earthquake epicentres from around the world, often from remote and distant locations, by correlating similar teleseismic wave forms between different seismograms. Her first papers appeared from 1926, on the reliability of various European observatories and the accuracy of such interpolations, contributing to the permanent record of important events that were then being published annually as the International Seismological Summary (ISS) from Kew, England. Reminiscing in old age she comments:

> As a result of many considerations it was found that while a group of stations did not allow travel times to be accurately determined, it made it possible to determine the slope of the travel time [its rate of change with distance]. In modern technology, it might be said the European stations were used as an array.

But she was not alone in her search for accuracy, for Bolt wrote tellingly of the early 1930s:

> She entered into a 'lively' correspondence with Harold Jeffreys [the leading seismological theorist of his day] during the period in which he and K.E. Bullen were calculating their famous seismological travel-time curves... In sharp contrast with the theoretical power of these two mathematicians, who reduced the published ISS sets of arrival-time readings of others, Lehmann, like Gutenberg, brought to bear the sharp observational insight provided by the seismographic patterns themselves.

It was this observational insight, a talent shared with Oldham, Mohorovičić and Gutenberg, but largely absent from the concerns of applied mathematicians working on the earth's interior, that would lead her, not them, to discover the inner core of the planet.

Background to discovery

As we have seen from the pioneering work of Oldham and Gutenberg, early seismological observations showed 'shadow' segments that interrupted P-wave arrivals at seismological stations beyond epicentral distances of around 120° of arc – a single core was proposed to explain such phenomena. The improved seismometers available to analysts such as Inge Lehmann in the 1920s and 1930s led to the identification of additional waves in shadow zones between 105° and 142° that could not be explained by the Oldham core. These unidentified waves were of P type but were delayed several minutes after arrival times predicted by standard P wave travel-time curves for 0° to 105° of arc – they were given the code of P′ waves. Various ideas were floated to account for the unexpected onsets, notably that such seismic energy was due to diffraction of the seismic waves at the 142° boundary.

Lehmann's own explanation for these mysterious arrivals began with a visual analysis and interpretation of the various seismological records from a seismic event in the southern hemisphere winter of 1929 (Fig. 4.2). Initially known as the Buller earthquake it is nowadays known as the Murchison event, whose epicentre was on the

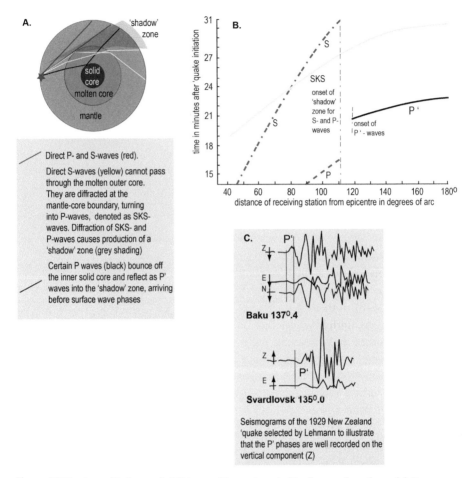

A.

'shadow' zone

solid core
molten core

mantle

Direct P- and S-waves (red).

Direct S-waves (yellow) cannot pass through the molten outer core. They are diffracted at the mantle-core boundary, turning into P-waves, denoted as SKS-waves. Diffraction of SKS- and P-waves causes production of a 'shadow' zone (grey shading)

Certain P waves (black) bounce off the inner solid core and reflect as P' waves into the 'shadow' zone, arriving before surface wave phases

B.

time in minutes after 'quake initiation

S

SKS

onset of 'shadow' zone for S- and P-waves

onset of P'- waves

P'

S

P

distance of receiving station from epicentre in degrees of arc

C.

Z

E
N

Baku 137⁰.4

Z

E

P'

Svardlovsk 135⁰.0

Seismograms of the 1929 New Zealand 'quake selected by Lehmann to illustrate that the P' phases are well recorded on the vertical component (Z)

Figure 4.2 Versions of Lehmann's 1936 opus illustrating: A. Her favoured earth model; B. Calculated travel times for that model assuming constant wave velocities in each layer: 10 km/s in the mantle, 8 km/s in the liquid outer core and 8.6 km/s in the solid inner core. C. Two of her illustrative seismograms from the 'shadow' zone. They show the P' phases well recorded on the vertical component (Z-direction) for two stations at arc distances in the 130°s. The subsequent high-amplitude wave traces are surface-wave arrivals from the same event.

northeast of South Island, New Zealand, its teleseismic traces recorded by ever-alert seismographic stations in the Russian and north European antipodes. The event was a big one, with a magnitude estimated as $M_s = 7.8$, one of the most powerful ever recorded along the seismogenic Alpine Fault that traverses northern New Zealand. It was triggered along the slip zone of a reverse (thrust) fault, probably of several tens of kilometres length (the mountainous terrain prevented exact measurements) with a typical vertical displacement of about 4.5 m, as seen where it cut across level highway surfaces. The reader might recall that both Oldham and Mohorovičić's earthquake faults were also thrust faults.

As we have seen, Lehmann was in an excellent situation to analyse the *P'* seismic wave phases from the event, not only because of her observational skills and natural curiosity but because the Danish seismographic network was located at a large epicentral distance from such energetic earthquake sources in the South Pacific. She could clearly see the *P'* waves on these and other seismographs as indicated by the arrows in Figure 4.2C. She wrote much later in a reminiscence, *Seismology in the Days of Old*, that:

> Among the phases of interest was *P'*, because of the longitudinal waves through the core of the earth. The rays are bent when they leave the mantle and enter the core, in which the velocity is much smaller. Thus the time curve has two branches. The first wave through the core (the one with the smallest angle of incidence) emerges at the surface of the earth at considerably greater epicentral distance and later than the wave that just touches the core. When the angle of incidence increases, the time curve runs backward until it stops at about 143° epicentral distance and runs forward again. Both branches of the time curve are indicated by European observations of the…New Zealand earthquake. The upper branch had not been indicated in Gutenberg's time curves and does not seem to have been observed before.

> At other distances, some *P'* observations were found that had not been explained. If the earth simply consisted of a hard mantle surrounding a fluid or soft core, we could not have observations recorded between 102°, where the direct *P* curve ended, and 143°, the smallest epicentral distance for *P'*.

> …Evidently, there was a reflection of the waves in the interior of the earth that caused them to emerge at a shorter epicentral distance. It was shown in a simple example how this could happen. I considered a globe in which a hard mantle surrounded a softer core, the radius of which I took to be five ninths of the surrounding sphere. The velocity of the longitudinal waves was 10 km/s in the mantle and 8 km/s in the core. It was then a simple matter to calculate the time curves arising from an earthquake that took place at the surface of the globe.…

> No rays emerged at epicentral distances between 112° and 154°. I then placed a smaller core inside the first core and let the velocity in it be larger so that a reflection would occur when the rays through the larger core met it. After a choice of velocities in the inner core was made, a time curve was

obtained, part of which appeared in the interval where there had not been any rays before....

Gutenberg accepted the idea. He and Charles Richter...placed a small core inside the earth and adjusted the radius of this small core until the calculated time curves agreed with the waves observed.

Lehmann summarized her results from the Murchison earthquake modestly in her 1936 paper:

It cannot be maintained that the interpretation here given is correct since the data are quite insufficient and complications arise from the fact that small shocks have occurred immediately before the main shock. However, the interpretation seems possible, and the assumption of the existence of an inner core is, at least, not contradicted by the observations; these are, perhaps, more easily explained on this assumption.

I hope that the suggestions here made may be considered by other investigators, and that suitable material may be found for studies of the P' curves. The question of the existence of the inner core cannot, however, be regarded solely from a seismological point of view.

Bolt attributed the strength of Lehmann's arguments to her ability to simplify the problem to hand, initially a two-shell earth model à la Oldham, with constant seismic P velocities (10 km s^{-1}) in the mantle and the core (8 km s^{-1}) and with seismic rays as chords rather than arcs. Then with a small central core transmitting a constant P velocity of 10 km s^{-1} she showed how a reasonable radius (1400 km) could predict a travel-time curve for the P3' core waves (Fig. 4.2). Bolt adds:

In effect, Lehmann proved an existence theorem: namely, a plausible tripartite earth structure could be found that explained the main features of the observed core waves. However, she did not go on to solve the inverse problem; that is, having proved the existence she did not use her measurements of travel times to estimate statistically the inner core parameters that satisfied them within the measurement uncertainties. This final step was done first two years later by Gutenberg and Richter (1938) who inferred an inner core radius of about 1200 km and a mean inner core P velocity of 11.2 km s^{-1}. They argued, however, that: 'Both observed amplitudes and travel times, suggest a rapid but continuous increase in velocity in a particular narrow range ... within the core rather than the discontinuity. Moreover, no reflective waves have been found, such as would correspond to a discontinuity.'

In 1939, Harold Jeffreys, after first giving a crucial test indicating decisively that the diffraction interpretation for the P3' phases was unacceptable, adopted the Lehmann inner-core hypothesis and obtained a satisfactory numerical inversion of his PKP travel-time dataset. But, unlike Gutenberg and Richter, he retained Lehmann's sharp boundary.

…It was not until 1962 that direct new evidence supporting Lehmann's sharp boundary was advanced, and not until 1970 that high-angle reflections (PKiKP) of seismic P waves incident on the inner core were observed unequivocally on seismograms.

In 1986 Lehmann wrote to Bolt concerning a special symposium at the American Geophysical Union Annual Meeting held in her honour:

I was, of course, aware that it was 50 years since I discovered the [inner] core but I did not pay much attention to the fact. I see now that I will have to take the anniversary more seriously… Fifty years ago, seismic instrumentation and interpretation had developed so far that the existence of an inner core could be established. Since then, great efforts have been made to determine the size and the constitution of the core, but no complete picture has been obtained and may not be obtainable by means of available observations. I understand that at this meeting all the more important results will be reported and discussed. This will undoubtedly be of great value and will be an event that will give much pleasure to those who are able to take part in it. I send my best wishes for a successful meeting.

Since her death in 1993 at the advanced age of 105 there have been many visual tributes to this remarkable scientist, two of which are reproduced in Figure 4.3.

Figure 4.3 Two rather different memorials to Inge Lehmann. A. The global Google headline of a few years ago celebrates a birth date anniversary. B. The city of Copenhagen celebrates its foremost female scientist in a more traditional vein, albeit with a splendid modernist installation on the pavement alongside other illustrious forebears of Danish science. The memorial, designed by Elisabeth Toubro, is installed on Frue Plads outside the University of Copenhagen. Wikimedia Commons.

PART 2

Drifting Stuff: Settings

Alfred Wegener (1880–1930) and Patrick Blackett (1897–1974)

Early Modern astronomers discovered our planet's sun-centred external motion, but recognition of its intrinsic internal mobility had to wait much longer. Despite pioneering studies using earthquake waves, such as the discoveries of the early twentieth century discussed in Part 1, there was little notion abroad that the planet might have an internal 'driving engine' for tectonics.

One general unifying theory appealed to a cooling and shrinking planet whose subsiding and crumpled surface crust was witness to episodic subsidence of former 'land bridges' as former continents foundered, but many thought that the continents and oceans were permanent. Little was then generally known about the nature of the sea floor and many nineteenth-century geologists were unfamiliar with observations and theories of Charles Darwin on the geology and subsidence of the basaltic Pacific Ocean floor with its sprinkling of both active and extinct volcanic islands. Darwin never approved of the *ad hoc* use of now-vanished continental 'land-bridges' to explain faunal and floral distributions – his frequent correspondence on the subject to such as Charles Lyell and Joseph Hooker reveal that to him the very words were like the proverbial red rag to a bull.

Both men featured in Part 2, Alfred Wegener and Patrick Blackett, were 'interconnectors' – oblivious of traditional academic subdivisions, conscious only of the unity given by application of the laws of mathematical physics to natural phenomena, whether atomic, meteorological, geophysical or geological. Wegener's continental drift theory was never in fact proved wrong, contrary to many authors' opinions. It was just that a plausible mechanism was lacking until in the later 1920s mantle convection was deemed possible and, 40 years later still, the possibility that lithospheric plates could slide about willy-nilly on Gutenberg's low-velocity zone. As noted previously, Wegener knew about Gutenberg's work and specifically highlighted its importance in the fourth and last revision of his book in 1928.

We may liken this problem of causation to Patrick Blackett's 'failed experiment' to determine the origins of the earth's magnetic field as due to planetary rotation, but whose apparatus nevertheless provided the means to determine ancient latitudes and magnetic pole positions. That a planetary magnetic field existed was beyond doubt since Elizabethan times and the studies of William Gilbert, but the actual mechanism for its generation had to wait until Walter Elsasser and Edward Bullard's work on dynamo theory came to fruition in the late 1940s, just as Blackett's research groups in Manchester and London began to grapple with the problems of accurately determining the often weak imprint of rock magnetism. In doing so, with the help of Blackett's sensitive astatic magnetometer apparatus, they confirmed Wegenerian drift theory beyond reasonable doubt.

5

Alfred Wegener (1880–1930)
German meteorologist and polar explorer

Wegener when at Marburg, 1910, age 30. Image from Wikimedia Commons via the German Polar Institute.

He proposed the first coherent mobilist theory of the earth from geological and topographical evidence, particularly of *c*.300 million-year-old polar ice sheet deposits, frigid climate plant species and by the refitting of now-opposed continental margins. All this suggested progressive break-up of supercontinental Pangaea by continental drift, forming all our modern oceans and destruction of the Tethys ocean along the Alpine–Himalayan mountain belt (1915–1928). Together with W. Köppen he produced the first global reconstructions of past climatic regimes (1924).

An audacious lecture

On 6 January 1912, at a meeting of the German Geological Association (*Geologischen Vereinigung*) at the Senckenberg Museum, Frankfurt-am-Main, a young lecturer from the University of Marburg, Dr Alfred Wegener, rose confidently (or so one must imagine, given the man's general character) to his feet and walked over to the dais of the lecture room to give his talk. The fact that he was a meteorologist about to present a revolutionary new theory concerning the behaviour of the solid earth at a meeting of professional geologists would not have worried him one iota. He had spent his spare time over the best part of two years researching his new theory and had carried with him to Frankfurt a typed-up manuscript of his arguments and conclusions ready for publication in the German equivalent of *Nature* magazine.

We do not know whether the Frankfurt lecture room was crowded with eager geologists attracted by the generality of Wegener's title on the programme of talks – *Die Entstehung der Kontinente* (The Origin of the Continents) – or whether it was scarcely attended at all, with perhaps other talks of more prosaic aims in competition. At any rate we must imagine Wegener's intensely focused blue-grey eyes fixing individual members of the audience as he spoke with firm but also modest directness to his subject. We must also assume that if his introductory remarks were more or less as given in his subsequent publication in *Petermanns Geographische Mitteilungen* for 1912, they would have had at least a few of the audience on the edges of their seats in anticipation of the main part of the talk that followed. Here is what he had written, as translated by biographer Wolfgang Jacoby, who keeps the traits of his personal modesty in his translation (unfortunately missing in other English translations), in 2001:

> In the following, a first tentative attempt will be made to give a genetic interpretation of the principal features of the earth's surface, i.e. continents and ocean basins, by a single universal principle of horizontal mobility of the continents. Wherever we used to have ancient land connections sink into the depth of the oceans, we shall now assume rifting and drifting of continental rafts. The picture we obtain of the nature of the earth's 'rind' is a new and somewhat paradoxical one, but as will be shown it does not lack a physical foundation. Such a large number of surprising simplifications and inter-relationships become visible after only a preliminary scanning of the main geological and geophysical results, that for that reason alone I consider it justified, even necessary, to replace the old hypothesis of sunken continents by the new one, because it appears to be more successful. The inadequacy of the old hypothesis has been demonstrated by its antithesis of the permanence of the oceans. In spite of its broad foundation, I call the new idea a working hypothesis and I wish it to be looked at as such, at least until it has been possible to prove by astronomical positioning with undoubtable accuracy that the horizontal displacements continue to the present day. Also, it is not superfluous to point out that this is a first outline of the hypothesis. In a more detailed elaboration, it will probably be necessary to modify the hypothesis here and there.

In suggesting, however tentatively, that the continents were in sideways motion and that the oceans were formed as distinctive entities, Wegener was offering a full-frontal attack on the traditional scenario of up-and-down tectonics and the notion of the oceans being largely the site of former continental crustal 'land-bridges' that had subsided and sunk without trace, like the legendary 'Atlantis'. His reference to the future science of GPS altimetry, his 'astronomical positioning', is a noteworthy demonstration of his perspicacity in this regard.

Early life and education

Wegener was a Berliner, born in November 1880, the youngest of five. His mother was Anna Schwarz, a great influence in his life. His father, Richard, had wide interests outside his day job teaching Latin and Greek at the prestigious medieval foundation of the Berlin Gymnasium at the Greyfriars Monastery. He was clergyman, intellectual, poet, and director of a school for 'orphans of worthy professionals'. Although religious, Richard Wegener was not a dogmatist and despite later inevitable disagreements between son and father, Alfred was left in charge of his own destiny and widening interests in the world about him as he grew older. Together with his brothers and sisters the Wegener children were allowed great freedom outdoors, actively encouraged by their parents, especially his mother, around the family summer home at Rheinsberg northeast of Berlin. This was near the 'lake district' and wild heaths of northern Brandenburg, famous for its association with the liberality of Frederick the Great's 'Rheinsberg Period' whilst Crown Prince in the early 1730s.

Wegener attended high school at the Köllnisches Gymnasium in Berlin, graduating top of his class. It was a modern establishment with less classics (though still with a hefty wallop of Latin) and a higher level of natural science, modern languages and mathematics than in the more traditional schools of the German education system at the time. During and after his high school career he developed into a natural 'interconnector' of both ideas and people. He became a driven empiricist, a collector of experimental measurements and observations, the consequences of which, after critical review, he unhesitatingly accepted, regardless of any initial positive or negative preference. He was intellectually fearless and completely unafraid of reasoned debate and argument. Such a personality could not have been bettered for subsequent entry into the controversies and animosities involved in global tectonic geology.

After graduation he studied physics, meteorology and astronomy to doctoral levels in Berlin, Heidelberg and Innsbruck. While in Switzerland in 1902–1903 he was an assistant at the soon-to-be-commissioned *Urania* astronomical observatory in Zurich. His Ph.D. in astronomy in 1905 was based on a dissertation written under the supervision of Julius Bauschinger at what is today the Humboldt University of Berlin.

Postdoctoral days: meteorology and Greenland

Wegener did not pursue a career in astronomy – his intellectual passion at the time was the pursuit of one branch of Lord Rayleigh's 'outdoor physics' – the nascent discipline of physical meteorology. Here the aeroplane, telegraph and wireless had begun to foster rapid advances in storm tracking and forecasting. His professional life thereafter also included strenuous fieldwork on long expeditions to remote polar territories in search of evidence for what became known as the jet streams or polar fronts of the high latitudes.

His first introduction to cutting-edge meteorology came in 1905 with his elder brother, Kurt, who also had an interest in meteorology and, later, polar research. They were hired as research assistants at the newly established Aeronautical Research Observatory at Lindenburgh near Beeskow, a rural area roughly 50 km southeast of Berlin, with Frankfurt-am-Oder to the east. The institute had formerly been in the capital itself but the latest trends in atmospheric research required kites and balloons to take measurements – activities inimical to safety and security in built-up cities. The observatory was inaugurated by Kaiser Wilhelm himself in October 1905 under the leadership of Richard Asswann who set up a Germany-wide network of five aeronautical observatories by 1907. These provided essential atmospheric data for the new and rapidly growing exponents of powered flight. Over their two years at the Observatory the Wegener brothers worked under Otto Tetens who had been developing a pilot balloon at the observatory whose purpose was to determine the characteristics of high-altitude winds. The two brothers must have flown many ascents in Tetens' balloon before their breath-taking achievement of 5–7 April 1906 – a new world record for continuous balloon flight, remaining aloft for 52.5 hours.

Later in that same year, Wegener participated in an equally challenging and potentially life-threatening experience as meteorologist on the first of his four Greenland expeditions. This, the Denmark Expedition (named after the ship involved, but also primarily a Danish-led expedition), was led by L. Mylius-Erichsen and charged with mapping the coastline of north-eastern Greenland, between Capes Bridgman and Bismarck. It was also the expedition that the young Karl Terzaghi missed out on due to a climbing accident in Austria (Chapter 10).

During the expedition, Wegener established the first two meteorological stations in Greenland, near Danmarkshavn, where he launched kites and tethered balloons to make field measurements in polar airspace. Here the dangers of fieldwork in extreme terrains became vivid to him following the tragic deaths of Mylius-Erichsen and two other experienced polar veterans who went on a long and unplanned exploratory excursion with sled dogs during which they became hopelessly lost, eventually perishing through exhaustion and lack of food.

Marburg 1908–1910: Köppen, climate and romance

After his return from Greenland in 1908 and until the onset of World War I, Wegener was a lecturer in meteorology, applied astronomy and cosmic physics at the University of Marburg. He gained a reputation there for his ability to clearly and understandably explain the basic physics behind the apparent complexity of natural phenomena like weather and other topics that were featuring in research efforts at the time. A physics colleague, Hans Benndorf, wrote of this young tyro:

> With what ease he found his way through the most complicated work of the theoreticians, with what feeling for the important point! He would often, after a long pause for reflection, say 'I believe such and such' and most times he was right, as we would establish several days later after rigorous analysis.

His lectures, informed by an exact understanding and knowledge of the current scientific literature, formed the basis of what was to become the standard textbook in meteorology in German, first written in 1909/1910 as *Thermodynamics of the Atmosphere*, and in which he incorporated many results obtained on the Greenland expedition.

It was also at this time that Wegener became interested and fascinated with climate, memorably defined as 'long-term weather'. Just how long-term that could be, extending into geological realms, Wegener was soon to explore with the help and guidance of a man who became both his mentor and, eventually, his father-in-law and co-author of a classic book, the first account of the science of palaeoclimatology. That man was Wladimir Köppen, the foremost climatologist of his day, widely regarded on account of his global climate classifications based on a quantified subdivision of the major climate zones. Wegener had got in touch with Köppen as part of his wider interests in climate from a polar perspective and sought his advice on several issues. He began visits to him in Hamburg, visits whose scope grew to include Köppen's daughter, Else, with whom he had fallen in love – the intense scientific connection became a close family one, to the benefit of all concerned.

Köppen had a joint Russian/German background and during his long academic career moved from St Petersburg to Hamburg in 1875, and later to Graz in Austria in 1924. In Hamburg he set up and ran the German naval meteorological service, collecting global temperature data and analysing its distribution. He devoted himself to his climatic research, publishing his famous classification of climates first in 1884, a definitive version in 1918 and a final statement in 1936 at the advanced age of 90 years. He died in 1940 and in addition to his intellectual achievements was remembered and valued for his deeply held social beliefs in equality, his interest in nutrition and in the contribution of a universal language (Esperanto) to world peace.

Marburg 1910–1912: drift – gathering the evidence

It was late in 1910 that 'the spark that led to Wegener's revolutionary mobilist ideas', in the words of biographer Wolfgang Jacoby, was struck in the small university town of Marburg. Summer was long past, also the early autumn migration of the myriad of swifts that still today in summer screech over and under the arches of the numerous medieval bridges that span the River Lahn. The town was dominated by its medieval university and its lively population of academics and students. Some contemporary individuals also scaled future heights of fame, notably the poet and Nobel Prize-winning novelist Boris Pasternak. He studied philosophy there in 1912 and wrote memorably about the town itself, the Marburg neo-Kantians and a failed love affair with a cousin.

Amongst the remains of the Christmas celebrations of 1910, Wegener's roommate brought out his Christmas present – a beautiful, just-published world atlas. Else Köppen quotes from Wegener's letter to her the following month:

> My roommate ... got Andree's Handatlas (5th Edition 1910) as a Christmas gift. For hours we admired the magnificent maps. A thought occurred to me: ... Does not the east coast of South America fit the west coast of Africa as though they had been contiguous in the past? Even better is the fit seen on the bathymetric map of the Atlantic when comparing ... the break-off into the deep sea. I must follow this up.

Others had of course noted the match and commented on it in various ways from the seventeenth century onwards, most recently the American F.B. Taylor. Wegener was not then aware of any of these commentators as he enthusiastically carried on his 'follow-up'. He later acknowledged Taylor's 1910 attempt to explain the most recent mountain-building by large changes in planetary oblateness, but remarked astutely:

> It does not appear that Taylor has realized the immense consequences of such large displacements of continents. As he did not investigate the implications despite their contradiction of traditional views, his suggestion has probably been met mostly with scepticism. For my work this was not the starting point ... I learnt about it later.

Wegener's Marburg colleagues helped him greatly by searching the geological literature and bringing him references. As he wrote: 'The geologists recover everything I need and bring it to me, saving nine tenths of my time. Otherwise, it would have taken months...' His charismatic personality cast an intellectual spell over his colleagues as he gathered relevant facts from this literature search in Marburg. One such was the young Hans Cloos, later to achieve fame for his work on the tectonics of continental rifting and related volcanism. He writes of his first encounter with Wegener, probably in 1911:

> One day a man visited me whose fine features and penetrating blue-gray eyes I was unable to forget. He spun out an extremely strange train of thought about the structure of the earth and asked me whether I would be willing to help him with geological facts and concepts.

Cloos remained sceptical but constructive, and Wegener updated his paper with new results and arguments. Here he was, by qualification a doctor of astronomy, a practising professional meteorologist, university lecturer, budding climatologist and recent polar explorer to boot. Yet drift and its geological, climatic and evolutionary implications were to become more than just an academic hobby as he prepared himself for another polar expedition just two years hence. He freely shared the progress of this hobby/obsession with Köppen, the older man sceptical to some degree at first. In her memoir, Else Wegener quotes from a letter he wrote in 1911 to her father, who had advised his young protégé against allowing himself to be 'side-tracked' from the serious polar meteorology that should have been his primary interest. Wegener responded:

> I think you consider my original continent [Pangaea] more fantastic than it is, and you do not see that it is but an interpretation of observations. ... We must assume a land connection, e.g., between South America and Africa, which ended at a certain time. It may have happened in two ways: (1) by the collapse of a connecting continent or (2) by the widening of a great rift. ... The first one contradicts the modern concept of isostasy and our physical understanding. A continent cannot sink, for it is lighter than that on what it floats. Considering the second hypothesis, we find a wealth of simplifications making more sense in the whole geological history; should we not throw the old hypothesis overboard?

Wegener's arguments in the 1912 paper

The editor of *Petermanns Geographische Mitteilungen* (Geographical Communications) accepted Wegener's manuscript with no alterations and published it in three parts. Wegener begins with geophysical arguments, but firstly deals with the apparent 'fit' of continental margins (Fig. 5.1), stressing that it is not the modern coastline that matters but that of the shelf margin, quite obviously a component and integral part of the continental crust:

> Since we deal with the continental platforms as a whole (i.e. continents plus shelves) it is necessary to free ourselves from the traditional picture of the coast lines. They are affected by shallow flooding of the shelves, the 'transgressions'... The continental break usually coincides with the 200 m depth contour, but for the Norwegian, the Barents, and the Iceland–Faeroe shelves among others...depths of 200–300 m reaching even 500 m at their margins. The picture [i.e., the continental fits] is only slightly changed by including the shelves, most strongly around Britain, northern Siberia, the Far Indies, and Bering Strait; furthermore, New Guinea now appears to be connected with Australia.

He goes on to rubbish the contracting earth hypothesis and the existence of 'land bridges' as proselytized by Eduard Suess, and, recognizing the contrast in density

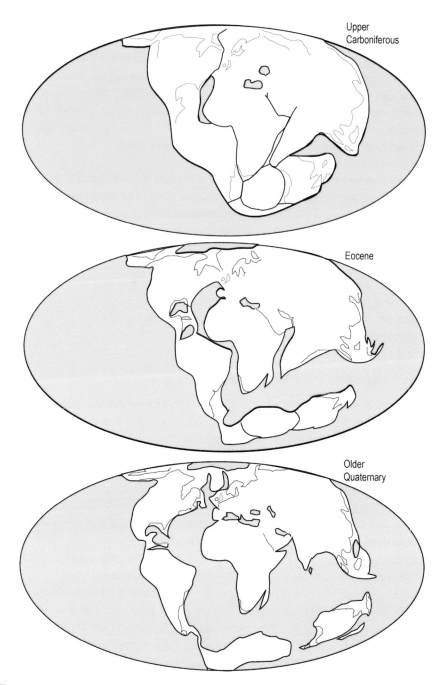

Figure 5.1 Simplified versions of the iconic maps of the changing continents illustrating Wegener's 'Displacement (*Verschiebungen*) Theory' that formed Figure 1 of the 3rd (1922) edition of his 'Die Entstehung der Kontinente und Ozeane', translated into English in 1924 under the title of *The Origin of Continents and Oceans*. I have added approximate dates for each.

and rock types making up the continents and oceans, affirmed the hydrostatic equilibrium provided by isostasy, something that would have had Darwin's, Oldham's and Mohorovičić's heads nodding in firm agreement.

He argued for a firm distinction between the brittle sial (granitic) of the continental crust and the oceanic crustal sima (basaltic), with melting beginning at greater depths to allow the horizontal passage of the former over the latter. Concerning melting:

> If we want to allow for horizontal displacements of [continental] blocks, we must assume a temperature at their bottom not far from the melting point. If we linearly extrapolated the geothermal gradient…we would expect the melting temperature at 48 km to 64 km… [but since] the main part of the earth is at a nearly constant temperature of about 3000 K; many investigators consider even lower temperatures likely. These arguments lead again to temperature values at 100 km depth not too far from the melting point of rocks.

This was pure Gutenberg before Gutenberg had even begun to think of his 'low velocity zone'. Taken as a whole, Wegener considers that the deeper sima must be in a state of plasticity such that, like loaded pitch, it flows on long timescales.

After a discussion of the deeply subsiding trough-like precursors to mountain belts containing great thicknesses of sedimentary materials, he concluded:

> In view of all the above, I conclude that mountain building continuing from the beginning and changing places, thickens the continental blocks at the expense of their lateral extent. For it is a unidirectional process: every compression causes thickening and decrease of surface area, but tension cannot cause the opposite; it rather disrupts the blocks.

This was another prescient anticipation of modern ideas concerning continental crustal shortening and extension. The end of this first section of his paper features a very short section, more of an *apologia*, concerning the possible causes of horizontal drift:

> The question, which forces cause the proposed horizontal displacements of the continents, is so manifest that I cannot quite bypass it although I consider it premature. It will be necessary first to exactly determine the reality and the nature of the displacements before we can hope to discover their causes. Here we can only attempt, above all, to prevent wrong ideas, rather than suggest something that can already claim to be correct.

> One may think of polar wandering as the cause; for any shift creates large additional centrifugal forces resulting in mass displacements. However, it will be shown in the last section that polar wandering is probably the consequence rather than the cause of the mass displacements.

I consider it more likely that the lunar tides of the earth's body are an essential cause. This appears to be supported by the preference for meridional rifting. This may also be the cause for a peculiarity, noticed often, that the continental shapes show a sharp thinning toward the poles.

This is most distinct in the region of the former south pole where the contours have not been disturbed by compression since rifting. The continents thin rather sharply, too, where the former north pole has to be assumed, at Bering Strait, as will be shown below; here, however, the contours do not seem to be unchanged, a consequence of the compression. Probably it will be wise for now, to consider the continental displacements the consequence of irregular currents inside the earth. In the future it will perhaps be possible to separate the essentially irregular, i.e. externally caused aspects, from the tendency toward an equilibrium in rotation. The old idea of a 'Polflucht' of the land which has recently been invoked by Taylor, in his work mentioned above, should belong to the latter class. The time is, however, not yet ripe for these questions.

Note the arresting phrase 'irregular currents inside the earth' *vis-à-vis* the future convection mechanism. As Wolfgang Jacoby notes, Wegener also envisioned a phenomenon akin to the modern concept of sea-floor spreading for explaining the growth of the ocean basins:

> ... we now seem able to explain the different ocean depths. Since for large areas we will have to assume isostatic compensation of the seafloor, the difference means that the seafloor we believe to be old is also denser than that believed to be young. Moreover, it seems undeniable that freshly exposed sima ... will for a long time maintain ... higher temperatures (perhaps 100°C in the uppermost 100 km on average) than old, largely cooled seafloor. ... The depth variation appears also to suggest that the Mid-Atlantic Ridge should be regarded as the zone in which the floor of the Atlantic, as it keeps spreading, is continuously tearing open and making space for fresh, relatively fluid and hot sima from depth.

This again is startling stuff to the modern reader aware of the great advances made in the early 1960s.

What of the direct empirical geological evidence for drift? Wegener is rather coy on this aspect, and one suspects that his hunt through the geological archives is unfinished and perhaps hasty in its execution. His first example is an observational and circumstantial view on modern rift valleys as extensional structures (formed by tectonic stretching) and wouldn't be amiss in any modern account of the initiation of sea-floor spreading:

I believe that all rifts should be interpreted…as fissures in the continental blocks and [of] incipient separation; they may be recent or former fissuring attempts that have failed because the driving forces vanished. The Rhine graben is probably of the latter kind for it originated as early as the Oligocene, simultaneously with the separation of North America from Europe.

He continues again in more observational mode with the relationships of geology in the western borders to the Atlantic (the Americas) with those to the east (western Europe, west Africa). He finds many intriguing similarities and suggestively fragmented geological boundaries, such as those bounding the coal-bearing strata of the Appalachians with the Variscan equivalents in northern Europe. None of these can be explained without drifting and separation. Concerning the relationship of the Andes with the opening of the South Atlantic he concludes that:

Since the Andes were folded simultaneously with the opening of the Atlantic the idea of a causal relationship seems obvious. The westward drifting American blocks would probably have encountered resistance by the very old and hard Pacific floor; this way the extensive shelf of the western continental margin with thick sediments would have been thrust into fold mountains.

Gondwanaland brings him a richer source of factual materials. He begins with fossil evidence:

The paleontological results leave no doubt that Australia had a direct land connection with India and South Africa/South America. This continent has been called 'Gondwana'; from the present position of the relicts, a huge size has to be ascribed to it. We must assume that Australia was part of it. Its separation from Africa and India seems to fall into the same time as that of their mutual separation, for during the Permian, immediately preceding the Triassic, the connection still existed, while during the Jurassic, immediately following the Triassic, the connection did not exist anymore.

He carries on with perhaps his strongest hand, that of the glaciation of the Gondwana southern continents, a long event around 300 million years ago in Permian-Carboniferous times and whose extent demanded the former conjoining of now separate parts:

Striking evidence for these ideas seems to come from the Permian glaciation (or Carboniferous, according to some researchers). Its traces have been found at many locations of the southern hemisphere, but not on the northern one. This Permian ice age until now presented a hopeless riddle

Figure 5.2 The classic ice-scoured pavement formed when the South African part of polar Gondwana was overrun by ice sheets from the north in Permo-Carboniferous times. Traces of the overlying Dwyaka tillite (ground moraine of the ice sheet) may also be seen. Sunglasses for scale. Source: AGU Blogosphere 9 February 2018 posted by Evelyn Mervine.

for palaeogeography. The undoubtable ground moraines of an extensive ice cap on typically striated basement are found in Australia, South Africa [see Figure 5.2], South America, and particularly in eastern India.

Such a conjoining would solve a great paradox, that of:

> …the present configuration of the land masses with an ice cap of such an extent is totally impossible. For…this would place the optimum pole position at the centre of the Indian Ocean and the farthest glaciated regions would still have a latitude of only 30–35°…The Permian ice age thus poses an unsolvable problem to all models that do not dare to assume horizontal displacements of the continents…If we reconstruct the configuration during the Permian, all glaciated regions shift together 'concentrically' to the southern tip of Africa, and we have to place the south pole into the now very restricted glaciated region. This would take everything mysterious away from the phenomenon.

He ends his paper on rather inconclusive and controversial topics (evidence for present displacements, polar wobble and polar wandering) that return to the already obviously vexing question in Wegener's mind – the elephant in the room of his entire proposal – a viable mechanism for drift!

Regarding the reception of the 1912 paper, biographer Jacoby writes perceptively:

> In the early decades of the 20th Century geographical research was closer to geology than it is today and geographers were less inhibited by geotectonic

doctrines. The editor of PGM, Professor Langhans, accepted Wegener's manuscript for publication unabridged, in three parts. The article caused great excitement, and not all of it negative. Its arguments ought to have made objective geologists think; but first reactions were inevitably rather subjective. Many renowned geologists ridiculed the idea.

War and publication of *Die Entstehung der Kontinente und Ozeane*

After enduring the first overwintering on the Greenland ice cap in the winter of 1913 and lucky to be alive after some hair-raising incidents, on his return Wegener and Else Köppen were married later in 1913 and settled down to life together in Marburg. He resumed his university lectureship and set about both researching in earnest his newly collected polar meteorological data and preparing drafts for a book-length account of the new drift theory. His eldest daughter was born in the year of the outbreak of war in which her father, as an infantry reservist aged 34, was called up during the general mobilization of 2 August 1914. One of the hundreds of thousands advancing *en masse* by foot into Belgium, swinging towards Paris, the German enactment of the infamous Schlieffen Plan went badly awry and was halted after a sharp and decisive Franco-British counterattack in early September at the First Battle of the Marne. Perhaps fortunately for him, given the years of deadly trench warfare that lay ahead, Wegener received two wounds in action at this period, which led to his hospitalization. After convalescence at home in Marburg he was assigned non-combatant duties as a military meteorologist for the rest of the war.

Despite having to travel constantly between meteorological stations across Germany and into the front line (where he might have come across Beno Gutenberg, who was similarly employed) and in adjacent occupied countries, his wartime activity left him time to complete *Die Entstehung der Kontinente und Ozeane* (The Origin of Continents and Oceans), published in 1915. Wartime conditions and the post-war chaos of German society meant that there was little wider interest in *Die Entstehung*, but by a continuous effort of further reading, thought and revision over the next several years, Wegener was able to bring out a second edition and then a wholly revised third edition in 1922. It was this that first became translated into English in 1924 (as discovered by the young Lawrence Bragg, see below) so creating the most intense controversy amongst English-speaking geologists and geophysicists since Agassiz's theory of the ice ages some 80 years before.

Climates of the past

After the war, Wegener obtained a position as meteorologist to the German Naval Observatory (*Deutsche Seewarte*) and moved to Hamburg with Else and their two young daughters. In 1921 he was appointed to a senior lectureship at the University of Hamburg and the whole family moved in with Else's parents. Close cooperation developed between Wegener and his father-in-law as they worked together on their geological history of climate book and, of course, a deep consideration of its implications for unravelling the history of the distribution of the continents. Else wrote that her father first opposed the new mobilist ideas of his son-in-law, but as he became more involved, he became convinced that the idea provided a way through the 'labyrinth of palaeoclimates'. Wegener wrote of how he enjoyed:

> ...a daily exchange of ideas with Köppen, and I had the satisfaction that he, initially cool and doubtful, increasingly warmed to drift theory, and finally convinced himself that here the 'red thread' through the labyrinth of palaeoclimatology had been found... Several chapters were written in such close exchange of ideas that the bounds of intellectual ownership can no longer be decided.

Four years of deep study resulted in their jointly authored 1924 book *Die Klimate der geologischen Vorzeit* (The Climates of the Geological Past; Figure 5.3), the same year that the Wegeners moved to Graz in Austria where he had been offered a professorship in meteorology and geophysics. Colin Summerhayes, a reviewer of the 2015 English translation of *Die Klimate* writes that it:

> ...featured the first comprehensive suite of global palaeoclimatic maps (displaying the distributions of climate sensitive indicators) for the Carboniferous, Permian, Triassic, Jurassic, Cretaceous, Eocene, Miocene, and Pliocene plus Early Quaternary, all made without the benefit of palaeomagnetic observations. Salt and gypsum deposits were found to be common where such evaporites are found today, in the arid belts north and south of the equator. Cretaceous corals characterized the equatorial zone between the 30th parallels, just like today. Glacial indicators clustered around the poles. And coals occurred under temperate humid conditions, and in the humid tropics. In the 1930s, Alexander Du Toit enthused about Köppen and Wegener's concepts, which enabled him to predict where past climate zones were.

Summerhayes' review focuses on what was to become perhaps the strongest piece of evidence for post-Permian break-up and drift of the continents – the great southern hemisphere glaciations of Gondwanaland (Fig. 5.4).

A particularly welcome positive voice of support for drift theory had come from Milutin Milanković (see Chapter 12) who had recently, and for several years in the future, would be in close touch with both Wegener and Köppen on matters concerning the interpretation of palaeoclimates. He wrote: 'I am impressed by your splendid concise and concentrated presentation. I am easily convinced by such a material of facts woven into a common idea.'

Wladimir Köppen • Alfred Wegener

The Climates
of the Geological Past

Die Klimate
der geologischen Vorzeit

Reproduction and translation
Faksimile-Nachdruck und Übersetzung
J. Thiede • K. Lochte • A. Dummermuth (Eds./Hrsg.)

Borntraeger Science Publishers

Figure 5.3 Wegener and his father-in-law, Wladimir Köppen, juxtaposed on the front cover to the bilingual German/English version of *Climates of the Past* published in 2015, some 91 years after the German first edition.

Figure 5.4 Version of the palaeoclimate information summarized here for the late-Palaeozoic of supercontinental Pangaea from *Climates of the Past* that provided the most telling support for Wegener's whole project. It was also the first-ever palaeoclimate map drawn up on a pre-drift re-assembly. Gondwana-born Alexander Du Toit used its evidence mercilessly against the anti-mobilists – such evidence eventually proved incontrovertible.

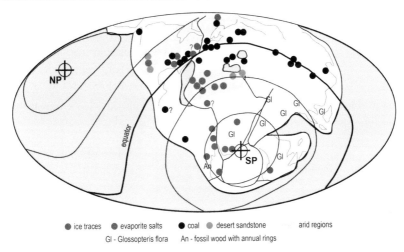

As we shall see in Chapter 12, Milanković had been following in the footsteps of James Croll some 50 years previously by appealing to orbital perturbations as the forcing mechanisms for climate change, by using his own advanced mathematical skills to compute values of past solar insolation. Köppen had convinced Milanković that the key to creating a glaciation was not the degree of winter cold but the change in insolation leading to increased duration and intensity of summer warmth.

The final edition of *Die Entstehung*

Wegener published his fourth and final version of *Die Entstehung* in 1927 (the English translation was in 1929), three years before departure for his final Greenland expedition that would lead to his early death at the age of 50. The book itself is a fitting memorial, for within its pages Wegener turns finally to the vexed problem of a mechanism for drift. He first discusses developments concerning the post-glacial 'rebound' of the Scandinavian land mass attributed to slow, inwards asthenosphere flow. As one of his key figures, he reproduced von Wolff's 1905 representation of the temperature distribution in the outer 120 km of the crust and upper mantle. This showed that for certain rates of temperature increase with depth there was an optimum region for melting 60–100 km down (Fig. 5.5). He wrote:

> [The] curves...are calculated with different assumptions about the crustal radium content. Additionally, two melting point curves...are plotted. Here, too, different curves are obtained according to the material assumed to be present. [The lowest] corresponds to the lowest conceivable fusion temperatures for the various depths. As shown by the knee of the temperature curves and the slope of the melting-point curves, there is an optimum region for melting at about 60–100 km down, and it is possible that here a molten layer is confined between two crystalline layers.

He is following here the work of Gutenberg on the upper mantle low velocity zone (Chapter 3) and writes in the accompanying text:

> It is natural to ask whether seismology might not provide an answer to the problem; it could do so if the molten state implied a low viscosity or

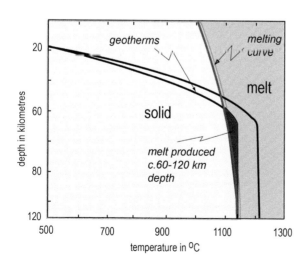

Figure 5.5 In his fourth and final edition (1928) of *Kontinente und Ozeane*, Wegener, possibly with Gutenberg's help, discovered this pioneering plot by von Wolff (1905) relating mantle melting to ambient temperature. The two temperature profiles (geotherms) intersect what Wegener referred to as 'the lowest possible fusion temperatures for the various depths'.

fluidity, since in a fluid medium there can be no propagation of transverse waves such as S-waves. It is generally believed nowadays, however, that any such material heated above the melting point and therefore molten or fused exists in an amorphous and glassy (and therefore solid) state. Nevertheless, seismology does give a pointer here. It can be shown that, making the most likely assumptions about the density of the material, its elastic resistance to deformation, which otherwise generally increases with depth, exhibits a discontinuity in this behaviour at about 70 km depth, and may even show a transitory reduction in resistance. Writers such as Gutenberg explain this by saying that, in all probability, at these depths the crystalline state is converted to the amorphous glassy one. Even if the glassy state should be considered as solid where short-period seismic waves are involved, it is nevertheless quite likely that it exhibits an appreciable degree of fluidity under forces which operate over geological timescales.

He replies directly to critics like Harold Jefferies who regarded the 'earth is as solid as steel':

> All that can be said with certainty is that the earth behaves as a solid elastic body when acted upon by short-period forces such as seismic waves, and there is no question of plastic flow here. However, under forces applied over geological timescales, the earth must behave as a fluid; for example, this is shown by the fact that its oblateness corresponds exactly to its period of rotation. But the critical point in time where elastic deformations merge into flow phenomena depends precisely on the viscosity coefficient...The problem of the viscosity of the layers lying below the continental blocks is closely dependent on whether the temperature of these layers exceeds the melting point or not. Although it is probable that the molten magmas can have a very high viscosity at very high pressures and therefore behave as a solid – the phenomena at such pressures are still unknown – yet all authors who support the idea of a molten layer tend towards the view that the viscosity of this layer is low enough to permit large displacements, convection currents in fact. Consideration of the radium content has produced quite new viewpoints on precisely this question...

This late mention of convection currents showed the direction in which Wegener was moving when it came to winkling out a viable mechanism for his theory.

Death on the icecap

Wegener returned to Greenland in the spring of 1930 as the leader of 21 other scientists and technicians. They were to systematically study the great ice cap and its climate from three observations posts. From the beginning things went badly. A very late spring melt prevented offloading of supplies from their vessel, leaving them a month behind schedule. A consequence was that Wegener estimated that the central field observatory was undersupplied to last out the winter and set out with another meteorologist and a dozen Greenlanders to resupply it by dog sled. They succeeded after a 40-day journey only to discover that their comrades did have enough supplies to last them through. Wegener and a Greenlander, one Rasmus Villumsen, set off back to the coast and were never seen alive again. An account by NASA tells it thus:

> It was not until May of the following year that a search party from the coast was able to locate Wegener's body. Villumsen had obviously buried Wegener with great care and respect, then presumably pressed on for the base camp, only to disappear into the white wilderness. Though a long, exhaustive search was made, the faithful Greenlander's body was never found...Wegener's brother and friends left his body as they found it and built an ice-block mausoleum over it. Later they erected a 20-foot iron cross to mark the site. All have long since vanished beneath the snow, inevitably to become part of the great glacier itself. It is a most fitting resting place for this remarkable man who devoted so much of his life to the study of that remnant of the last ice age and whose vision of moving continents provided the key to the mysteries of more.

The initial reception of Wegener's ideas

Alfred Wegener was still alive when Arthur Holmes read his celebrated paper to the Glasgow Geological Society in 1928 (see Chapter 7) in which he discussed radium-generated heat in the earth and advocated continental drift and oceanic formation by slow-moving mantle convection cells. Had Wegener known of its content and the vigour of Holmes' defence of the mobilist cause, he would doubtless have been delighted. Here was the answer to enthuse his supporters and dismay his critics – Holmes' convection model, though barely developed as a physical mechanism, was what he had sought all those long years since 1912.

Who were these supporters and critics? Naomi Oreske dealt with this question from an American viewpoint in fine measure in her 1999 *The Rejection of Continental Drift: Theory and Method in American Earth Science*. However, it is important to make clear that numerous prominent and talented geologists and other scientists worldwide (including a Nobel prize winner) supported the drift concept during the 16-year timespan of the four editions of Wegener's book (1912–1927) and in the decade following his death. The First World War had prevented the Anglosphere's access to German science (and *vice versa*) till 1919, but thereafter the list of supporters included some notable geologists – Holmes himself at Durham, Reginald Daly at Harvard, Beno Gutenberg at Caltech, Edward Bailey at the British Geological Survey, Émil Argand at Neuchâtel, Leon Collet at Geneva, Alexander du Toit at Cape Town and Felix Vening

Meinesz at Utrecht – a glittering array, all mobilists from the 1920s onwards. Later came Harry Hess and J. Tuzo Wilson, both exposed to the Wegener controversy during their university educations at Yale, Princeton and Toronto (see Chapter 16).

Physicists, meteorologists and mathematicians also joined the ranks of the mobilists. Most notable were two youngish Manchester professors. One was the Nobel prize-winning crystallographer, Lawrence Bragg, Professor of Physics and successor to Rutherford after his move to Cambridge. The other was the pioneering meteorologist and mathematician, Sidney Chapman, Professor of Mathematics. Their fascinating behind-the-scenes role was not recognized until an intriguing study by historian Ursula Marvin, *The British Reception of Alfred Wegener's Continental Drift Hypothesis*, of 1985. She wrote:

> Bragg's name was never formally associated with Wegener's theory until 1968, when a testimonial volume was published honouring Sydney Chapman, the mathematician, geophysicist, and meteorologist, who had been a Professor of Mathematics at Manchester in 1922. In that volume Bragg recalled walking the Derbyshire hills with Chapman who, one day, gave him an account of Wegener's theory of continental drift. The idea impressed Bragg so much that he said he could still remember the exact spot where Chapman began to talk about it. Bragg wrote 'I was so thrilled that I wrote to Wegener for an account of his theory, got it translated, and presented it to our Manchester Literary and Philosophical Society. The local geologists were furious; words cannot describe their utter scorn of anything so ridiculous as this theory, which has now proved so abundantly to be right.'

Chapman himself had learnt of Wegener's ideas from Norwegian geologists at Oslo during an academic conference some years previously. Marvin reckoned it was probably Bragg, who, after receiving his translation of Wegener's ideas behind his book, wrote the unsigned review headed as 'Wegener's displacement theory' in *Nature* in February 1922. In this way Bragg became responsible for kick-starting awareness of the Wegenerian revolution in Britain and widely elsewhere. But, by withholding his name as author, such approbation by a celebrated Nobel Laureate was somewhat reduced – a missed opportunity perhaps for the mobilist cause? The review began and ended thus:

> This book makes an immediate appeal to the physicists, but is meeting with strong opposition from a good many geologists. This opposition is to be expected, for the author replaces the whole theory of sunken continents, land bridges, and great changes of earth temperature by a displacement theory...

> The revolution in thought, if the theory is substantiated, may be expected to resemble the change in astronomical ideas at the time of Copernicus. It is to be hoped that an English edition will soon appear.

A third German edition duly appeared in 1922, which received its first English translation in 1924.

The objectors to the substantive part of Wegener's theory, the geological evidence, were themselves geologists and palaeontologists, many of them North Americans (the centrepiece of Oreskes' book), perhaps with a strong dose of the anti-German chauvinism that was so commonly expressed in both wartime Britain and the USA. There was also a wilful ignorance of developments in European geology, like the success of Alpine nappe tectonics as described by Argand and Collet. These brilliant field geologists had envisaged that the compressive driving force to form such structures came from the northwards drift of continental Africa relative to Europe and the closure of the intervening Tethys Ocean, as Wegener's palaeogeographic maps had indicated.

Many critics would also not accept a theory without a viable mechanism. These were the 'fixists' and 'verticalists' – true-believers in the permanence of the ocean basins, allowing only up-and-downs of rigid continental crustal fragments. This they explained by using Barrell's isostatic principles, but somehow forgot along the way that these required both a denser and a less-than-solid asthenosphere under their fixed and bobbing continents. In a way, such believers were betraying the philosophical basis of geology, denying the essential pieces of decisive field evidence that had led Wegener inexorably to his conclusion that lateral drifting had occurred – chiefly the polar glaciation of the southern part of his Pangaea supercontinent. As Edward Bailey (see Chapter 13) stressed in the 1930s, this evidence admitted only one solution, and Wegener had called right. Many other thoughtful geologists who contributed to the great debate of drift also felt the same way. Here are the succinct and penetrating comments of sedimentary geologist, Robert Rastall, again as researched by Marvin, concerning cause and effect, made in 1929 in the *Geological Magazine*:

> It is hardly too much to say that the present status of the [continental drift] controversy is that geological evidence continues to accumulate showing that lateral movement of continental masses has taken place, as is indeed admitted, either directly or tacitly, by many of the opponents of the Wegenerian theory, while mathematicians and cosmogonists continue to reiterate that such is impossible. Now the real meaning of this attitude is that the mathematicians and astronomers have not yet discovered a cause, which is not by any means the same thing as proving that there is no possible cause, a philosophical distinction which is very commonly disregarded. A similar state of affairs has existed several times before, even in some of the more important problems of quite modern geology.

Rastall went on to cite three examples of such important problems: first, the age of the earth as miscalculated by Kelvin's conductional cooling hypothesis in the face of sedimentary evidence for a very much greater longevity; second, the origin of ice ages, '...despite the irrefragable evidence that they do occur'; third, the evidence from Alpine nappes that hard sedimentary rocks can be '...folded up like a table cloth back from the wash...' despite its impossibility from the physical laws of the deformation of brittle materials.

Rastall's comments regarding the primacy of field evidence in any geological argument, leads us to the determined and lifetime opposition by the Cambridge mathematician and seismologist, Harold Jeffreys. He first made public his views on a drift mechanism in the autumn of 1922 at a meeting of the British Association where he pointed out, quite rightly, that the rotational forces proposed by Wegener were totally inadequate to drive the motion of continental masses. He returned to the topic at a meeting of the Royal Geographical Society in early 1923:

> My main complaint against the theory is that the physical causes that Wegener offers for the migration of continents are ridiculously inadequate. They are something like a millionth, or less, of what is required to produce the shearing of the continents.

So, it was 'no mechanism, no drift' – a direct challenge to geological logic of the kind analysed by Rastall and taken up by the efforts of people like Holmes, Daly and du Toit over the next decade. The former two made use of convection currents in their search for a mechanism whilst du Toit's huge assembly of additional and incontrovertible geological evidence was put together in his classic book, *Our Wandering Continents* (1937), initially inspired by the Permo-Carboniferous glacial deposits of his South African homeland. We can also include Beno Gutenberg to this list, for after his move to California in 1930, his repeated espousal of a low-velocity zone from detailed analysis of earthquakes records in his adopted country led him to even more strongly disagree with Jeffreys' conclusions concerning upper mantle rigidity. His new Californian school of earthquake seismology would lead the way in subsequent decades, making Jeffreys' models of an entirely solid mantle irrelevant.

6

Patrick Maynard Stuart Blackett (1897–1974)
English physicist

Blackett, probably taken in the mid- to late 1930s. Wikimedia Commons.

He enabled the detection and analysis of weak magnetic fields preserved in sedimentary rocks by the design and use of a sensitive astatic magnetometer (late 1940s) that could measure magnetic declinations which, through the work of associates and doctoral students, confirmed core dynamo theories for the generation of the planetary magnetic field and Wegener's theory of continental drift (1952–1961).

Introduction

It is almost stranger than fiction that a Nobel Prize-winning atomic scientist who had previously seen active service as a young naval officer in two notable sea battles of World War One, should in the following war establish the discipline of Operations Research and the adoption of the North Atlantic convoy system that finally defeated the Nazi U-boat menace in 1943–44. Further, in the post-war years he turned his interest to the origins of the natural magnetism of the earth's rocks, inventing a sensitive magnetometer to detect the direction and intensity of the weak magnetism present in many of these. That person was Patrick Blackett and this chapter is the story of his magnetic life and personality, together with his students and research associates. Together they kicked off entirely new avenues of research that revolutionized the nascent earth sciences from the early 1950s, confirming continental drift and the break-up of Wegener's Pangaean supercontinent.

Magnetic navigations

On our near-spherical globe the lines of latitude and longitude are impressed upon us in early geography and geometry classes as the chief coordinates whereby any position on earth may be determined. The discovery long ago that earth had a particular magnetic field introduced an additional navigational aid – the lines of magnetic force change their tilt (inclination) from pole to equator in a regular way given by simple trigonometry. Later, measurements of magnetic directions in young volcanic lavas showed alignment with the earth's field, introducing the phenomenon of rock magnetism. The idea came about that Wegener's proposal of continental drift could be tested by measuring such alignments in ancient continental lavas, but also in sediments and sedimentary rocks containing iron-bearing minerals whose ages were known from either radiometric dating or by the fossil content of interbedded sediments. If the continents had once been part of Pangaea, for example, and then moved their latitudinal positions as Wegener had suggested, then their inclinations before and after movement might be different. But what if instead the magnetic poles had migrated? Or perhaps both magnetic polar wandering *and* drift had occurred?

The background to all this is that our planet is thoroughly sideritic – iron is its most abundant element – courtesy of the 'Big Bang' as the universe began and the solar system nucleated. This abundance, with the element's exceptional intrinsic atomic properties, means that earth magnetism is a force that spans not only the whole of the solid planet, but also extends outwards into space to encircle it. The relevant properties of the iron atom, one of the transition series in the chemical periodic table, are due to a couple of unpaired outer electrons whose errant and unbalanced spinning set up vortex motions that create a field of magnetic force along their spin axes. The visual evidence for these force fields might be recollected by readers who in their youth were thrilled by that most visible of physics experiments – the beautifully symmetrical 2D pattern of iron filings scattered randomly on a sheet of paper and placed over a bar magnet – a demonstration of the axial dipolar field about magnetic north and south. The magnetism of rocks depends upon the presence of ferromagnetic iron oxide minerals, magnetite and haematite, substances which can take up and hold onto the magnetic field vectors of the planetary dipole picked up at their time of crystallization for hundreds of millions of years.

In the early 1970s my palaeomagnetic colleagues and students under Jim Briden at the University of Leeds were known affectionately, but with a degree of suspicion, as 'palaeomagicians'. The lively group, with John Piper and Bill Morris prominent, were trying to sort out the likely width of the ancient Iapetus Ocean using by-then routine palaeomagnetic techniques. We wondered how rocks could take up and hold the orientation of a past planetary magnetic field on such geological timescales in the face of possible chemical alteration or recrystallization. Perhaps we had subconsciously imbibed the forceful disregard of Harold Jeffreys, who in 1959 in the final edition of his *Earth*, wrote the following rather disdainful comments on the status of rock magnetism:

> When I last did a magnetic experiment (about 1909) we were warned against careless handling of permanent magnets, and the magnetism was liable to change without much carelessness. In studying the magnetism of rocks the specimen has to be broken off with a geological hammer and then carried to the laboratory. It is supposed that in the process its magnetism does not change to any important extent, and though I have often asked how this comes to be the case I have never received any answer.

William Gilbert: early modern magnetic experimenter and theorist

Any discussion of modern planetary magnetism must include an account of the extraordinary achievements of an Elizabethan scientist, William Gilbert, in explaining the causes and patterns of magnetic inclination – just as it would be impossible to discuss mechanics without reference to Isaac Newton's achievements just 90 or so years later. Gilbert was born in Colchester, Essex, in 1544, the son of minor gentry, making it to the royal courts of London through his medical training – Royal Physician to Elizabeth I and President of the Royal College of Surgeons. But his great private passion for more than 20 years was magnetism. In 1600 he published his *magnum opus* in Latin, *De Magnete* (On Magnetism), which contains the first proposal concerning the origin of the magnetic state of the planet. Gilbert's inquisitive and sceptical mind had become increasingly dismissive of Hellenistic (Aristotle's) science and its wider influence on contemporary religious dogma, specifically in its failure to undertake rigorous experiments with the aim of explaining natural phenomena. He wrote in *De Magnete*:

> We do not all quote the ancients and the Greeks as our supporters, for neither can paltry Greek argumentation demonstrate the truth more subtilly [sic] nor Greek terms more effectively, nor can both elucidate it better. Our doctrine of the loadstone is contradictory of most of the principles and axioms of the Greeks.

In this regard alone Gilbert was a very early and insistent proponent of empirical science – that done by observation and/or experiment and inductive reasoning. He was on a par with the giants of early modern astronomy, Copernicus, Kepler, Brahe and Galileo – and an easy match for Francis Bacon's better-known achievements in his more widely known *Novum Organum* of 1620. Yet Bacon had conducted few, if any, experiments of his own, unlike Gilbert, who had become fascinated by the origins of magnetism because of his interest in the voyages and discoveries of the great Elizabethan maritime explorers.

Gilbert begins *De Magnete* with an acknowledgement to such contemporaries who provided instruments, as his 'learned men') who 'have invented and published magnetic instruments and ready methods of observing, necessary for mariners and those who make long voyages…' He specifically mentions, among others, one Robert Norman, a London compass-maker who had long been frustrated by the time and effort he had spent on balancing his finest vertically balanced compass needles to correct a seemingly constant and steep downward tilt. Such a concern led Gilbert to infer the existence of deep, whole-planet magnetism that attracted magnetic needles thus. He designed experimental apparatus and set about a long series of experiments to investigate his inference further.

Using a procedure previously introduced by Norman to find the local tilt of his compass needles, he investigated the interaction of tiny magnetic wires impaled in cork and floating in a tray of water above specially forged and tooled spherical magnets to mimic the earth. He found a fundamental unity in the polar 'attraction' of the magnetic spheres– a systematic and gradual change from zero tilt around their equator (horizontal wire) that gradually increased with latitude to 90° (vertical wire) at the poles (Fig. 6.1). This explained London's high dip of around 70° at its latitude of 52 °N and enabled the geometrical calculation of latitude anywhere on the globe. In this way Gilbert established that the earth is a wholly magnetic planet, with its field of force stretching out over oceans and continents alike – he was both an observational and a theoretical scientist.

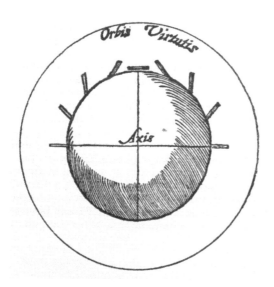

Figure 6.1 William Gilbert's illustration from *De Magnete* of his experimental results concerning the variation of magnetic inclination with latitude (somewhat misleadingly the diagram has the equatorial axis vertical). The relationship is nowadays central to the science of rock magnetism and is expressed in terms of a trigonometric expression – the tangent of the angle of magnetic inclination is twice that of the angle of latitude.

One feature of the earth's magnetic field that Gilbert did not investigate was the puzzling variation, particularly at high latitudes, of the vertical magnetic north pole when measurements over scores of years apart and from fixed locations showed variability of up to 13° from the earth's rotational axis. However, in Gilbert's time there was no search for systematic trends in such aberrations, and so he made the simplest assumption possible, that the magnetic and rotational axes of the earth coincided. Centuries later this variation was found to be due to gradual and systematic movements of magnetic north – a phenomenon now known as the secular magnetic variation.

Patrick Blackett: life, physics and operational research

Blackett was described by his *Nature* obituarist, Edward Bullard, himself a remarkably wide-ranging physicist-turned-geoscientist, as 'the most versatile and the best loved physicist of his generation...he was wonderfully intelligent, charming, fun to be with, dignified and handsome.'

How this reputation came about is paralleled, but not surpassed, by that of Harold Urey, whom we shall meet later in this book – they were both Nobel Prize winners, Urey for chemistry in 1934 and Blackett for physics in 1948.

The son of a well-off stockbroker, Blackett was one of the generation of young men and women across Europe and beyond, whose 'normal' lives were brutally interrupted by participation, often during active service but also as scientific advisors, in one or both world wars. In Blackett's case, before World War I, he had already opted for a career as an officer in Britain's Royal Navy following graduation from the naval college in Devon. In 1914 as war began, he was enlisted into active service afloat as a midshipman. He served in ships that fought in the major naval battles of the Falkland Islands and Jutland. Thereafter he was a junior lieutenant in patrols on lighter craft in the English Channel. At war's end, like many members of the surviving British officer corps (as Ralph Bagnold in Chapter 11), the government offered him a place at university.

Blackett's maritime career in the Royal Navy, though short, was rich in potential practical application and the benefits of applied physics that had stimulated his own first technical innovations in ballistics and improved gunsights – his decision after the war to study mathematics and physics at Cambridge was not unexpected. What was unexpected was his parallel discovery of the intellectual delights of free speech, multidisciplinary exchanges and friendships that developed in the post-war Cambridge colleges – a contrast to the rather restricted and hierarchical talk that he was used to in naval wardrooms. He met free-thinking socialists and novelists who stimulated his already developed independent mind and jump-started his future stances to the democratic left of the political spectrum. At work on his courses, he met members of one of the most brilliant groups of physicists in the world whom Ernest Rutherford had attracted to the Cavendish Laboratory.

After two years of undergraduate study he decided to resign his naval commission and join the Cavendish group as a researcher under the great man. He thrived in the exciting atmosphere generated by the informal and gregarious Rutherford (plenty of golf and tennis) and produced spectacular results from long hours examining images from the cloud chamber invented by Charles Wilson in 1911. This visualized by photographic exposure the vapour trails resulting from the ionization of gases by incoming radiation, and was specially modified by Blackett. It enabled him to record in 1924 the first visualization of the transformation by radiation of one element to another, in this case nitrogen which, by proton ejection, turned into one of the two stable isotopes of oxygen – we shall meet such isotopes again when we consider Harold Urey's achievements in Chapter 13.

As the first modern alchemist and pioneer of cosmic ray research, Blackett went on, with Giuseppe Occhialini, in the early 1930s, to invent an ingenious automated cloud chamber whose Geiger counters detected passage of high-energy, electrically charged cosmic ray particles (originally protons and alpha particles) which then triggered automatic exposure ('photographing themselves' as he put it in his Nobel lecture) and of the existence both of 'normal' negative electrons but also of positively charged electrons. In his own words:

> This fact of the rough equality of numbers of positive and negative electrons, and the certainty that the former do not exist as a normal constituent of matter on the earth, led us inevitably to conclude that the positive electrons were born together in collision processes initiated by high-energy cosmic rays. The energy required to produce such a pair is found from Einstein's famous equation ($E=mc^2$). So was demonstrated experimentally for the first time the transformation of radiation into matter.

In 1933 Blackett left Cambridge with his experimental kit and set up successive highly productive and diverse research groups, first at Birkbeck College, London and then at Manchester, where he succeeded W.L. Bragg of X-ray crystallography fame (see also Chapter 5). Both before and during World War II he supported the armed forces of his country in their fight against Nazism by being prominent on committees formed to specifically improve the direction of armed conflicts of every kind. To this end he created the field of Operational Research which had one overriding objective in the fields of military strategy and tactics – the replacement of emotional decision-making with rationality. He was successful in his efforts to change the course of the war in the North Atlantic against the German U-Boat menace by the adoption of convoy tactics and by the rapid and directed use of reliable intelligence information in guiding counterattack by long-range aircraft. However, he was not successful in his total opposition to Bomber Command's Air Marshal 'Butcher' Harris's relentless policy of 'area' (indiscriminate) bombing of German cities (joined after 1943 by the USAF flying from East Anglian bases) as an achievable war aim – it was an emotional policy, not rational – with great loss of life on both sides.

From cosmic rays to rock magnetism

During his post-war years at Manchester and his subsequent move to Imperial College, Blackett's formidably fertile mind turned earthwards and to the distant legacy left by Gilbert's *De Magnete* published, as we have seen, some 450 years previously. He wanted to know the true origins of the planetary magnetic force and to understand and accurately measure rock magnetism – resolving the origins of magnetic field reversals and of the major controversial geological issues left in limbo by Wegener's theory of continental drift. To this end he designed and constructed a special astatic magnetometer whose workings involved suspended rigid pairs of opposed magnets that responded to the magnetic field of an orientated rock specimen but not to the static magnetism of the geomagnetic field or of any laboratory disturbance. Concerning the reasons behind the development of this general interest he wrote that in the late 1940s:

> Under the stimulus of Babcock's discovery of magnetic stars I had previously interested myself in the early speculations on possible origins of the magnetic fields of astronomical bodies...of the origin of the field of the earth and the sun in some fundamental relationship between magnetism and rotation. Incidentally, my interest … was originally due to an attempt to calculate the deflection of a cosmic ray in traversing the galaxy. So the steps from my work on cosmic rays to that on rock magnetism were few and short.

He had in fact abandoned his efforts to establish any relationship between magnetism and planetary rotation by 1949, his carefully crafted astatic magnetometer failing to detect any evidence for it in experiments using suspended and rotating gold cylinders. More importantly, another new idea *was* testworthy. It was proposed to Blackett by Edward Bullard and concerned his new core dynamo theory for the origin of the field. Bullard had deduced that this would cause a downwards increase in field strength through the earth. Blackett passed on this proposal to his bright and mercurial graduate student, Keith Runcorn, who, with the help of fellow researchers at Manchester, A.C. Benson, A.F. Moore and D.H. Griffiths, made underground magnetic measurements in deep mines in the northern coalfields of England. These most definitely indicated a Bullardian increase, the affirmative result published in 1951.

Despite titling his own long paper on the failed rotation theory in 1952 as a 'negative experiment', the astatic magnetometer featured strongly in it and proved itself highly sensitive and quite capable of detecting weak magnetic remanence in sedimentary rocks. Blackett had by then fully accepted Bullard's dynamo theory (also proposed independently by Walter Elsasser in the United States). Here he is writing in his usual easy and fluent style in his classic book *Lectures on Rock Magnetism* (from his Weizmann Memorial Lectures of 1954 at Rehoveth, Israel) that, although there had been much theorizing on the subject:

The most plausible attributes the earth's magnetic field to electric currents circulating in the liquid core…due to some kind of electromagnetic dynamo driven essentially by a heat engine deriving its energy from the radioactivity of the material of the core.

He goes on, however, to stress that the lack of information concerning the actual core material's physical properties and the inherent difficulty of quantifying the processes involved made it impossible to deduce magnetic behaviour in detail. In his prescient words:

For many years all that one can hope to do by such theories is to understand in outline how the earth behaves magnetically as it does. So more facts about the magnetism of the earth become an essential basis for successful theorizing.

This was his clarion call for the need of a massive international campaign of continental and oceanic magnetic data collection designed to elucidate and explore the origins and significance of field reversals and drift theory. One hears echoes of William Gilbert yet again – he was asking his research group and others around the world for new voyages of navigation and discovery in those brave post-war years of recovery. Spirits had begun to rise again in Britain at this time – my later schoolboy stamp collection featured 1951 as the year of the 'Festival of Britain', celebrating the 100 years that had passed since Prince Albert's Great Exhibition and of his idea of founding the scientific Royal College in London, of which Blackett was now a part in its new guise as Imperial College.

Magnetism and drift

Regarding magnetic reversals, Blackett wrote in the Weizmann *Lectures*:

If twenty years ago one had speculated about the type of results likely to be obtained from the study of the distant past history of the earth's magnetic field, one would reasonably have expected to find that it had remained in roughly the same direction as at the present epoch, but one would not have been surprised to find that it had been much bigger (or much smaller) in magnitude in the past. As has so often happened in the history of physics, the experimental evidence proved of a quite unexpected kind. A large number of rocks of widely differing types and ages were found to be magnetized in roughly the opposite direction to that of the present field…

…if this remarkable result is a true deduction from the facts, then it must dominate all future theories of the origins of the earth's field and, incidentally, teach us, however indirectly, a great deal about the nature of the earth's core and of its complicated motions. For a three-dimensional dynamo in a rotating sphere of conducting liquid, which every so often suddenly reverses the direction of its electrical output, must have some very special characteristics. The postulated sudden field reversals would provide a strait jacket of assumed experimental fact into which all future theories of origin of the earth's magnetic field must fit.

Regarding continental drift, in subsequent years and into the 1960s and beyond, Blackett's team of Manchester and Imperial College researchers dispatched themselves to all the continents in order to gather data on rock magnetic vectors in order to achieve Blackett's goals, but especially to test drift theory. Noisy water-cooled saws and drills (Fig. 6.2) echoed through the quiet places of the world: Torridonian Assynt in the Scottish Northwest Highlands; Arizonan desert canyons; snow-covered Icelandic volcanic tablelands and volcanoes (by J. Hospers, who had first proved magnetic reversal here) and the gentle pastoral undulations founded upon English, Irish and

Figure 6.2 A. Obtaining cylindrical, orientated palaeomagnetic sample cores in hard rock by a portable, water-cooled drill can be wet, cold, dirty and dangerous, as demonstrated here in Jurassic red bed sandstones of Tethyan margin origins by Doyle Watts and Lin Jinlu during the Sino-British Tibetan Geotraverse of 1985. B. By way of contrast, Steve Salyards has hand-drilled out this core from softish Pleistocene red bed siltstones in the Rio Grande rift in the 1990s to determine their magnetic polarity.

Scottish Mesozoic sandstones. Thousands of samples, many from other laboratories and countries, notably the USA, who joined the magnetic chase, provided the two key components of the ancient field – Gilbert's magnetic inclination (to define latitude) and declination (to define the magnetic pole).

Keith Runcorn became a researcher at Cambridge in 1950 after the startling success of his Ph.D. research under Blackett at Manchester, and here he inspired, somewhat chaotically by all accounts, his own talented research group, including K.M. Creer, E. Irving, J.H. Parry and D.W. Collinson. They, and their own students in turn, further developed the astatic magnetometer and the details of the complicated data reduction and statistical corrections (courtesy of R.A. Fisher) necessary to isolate the 'primary' ferromagnetic remanence acquired in once-deposited but now long-buried and variably recrystallized sedimentary strata. Such magnetism is biochemical in its origins – the product of magnetite microcrystals precipitated by biomineralizing bacteria and others. These use the reduction and oxidation of iron to produce their own metabolic energy, at the same time aligning themselves along the magnetic dipole field.

Common igneous rocks like basalt lavas and dolerite sills were also sampled. These possess larger crystals of magnetite which gain strong (thermo-remanent) magnetism after passing through their Curie points following crystallization from cooling magma. This critical temperature, discovered by Pierre Curie during his Ph.D. research, is around 750°C. Below this molecular kinetics allows the preservation of the local magnetic vector.

Prominent in the early 1950s amongst Runcorn's graduate students was Edward Irving. Afterwards, at the Australian National University, he was encouraged and inspired by J.C. Jaeger to develop his magnetic work after negative Ph.D. experiences at Cambridge. Together with Runcorn and Creer he became one of the most prominent individuals in the search for evidence of both polar wandering (as long-term movements of magnetic north became known) and continental drift. He not only made many magnetic determinations but also added canny palaeoclimate nous to his intellectual kitbag. He had read the sedimentary literature (and presumably Wegener and Köppen) closely and was able to check his magnetic vectors against hard geological evidence for palaeoclimates gained from sedimentary rocks in the stratigraphic record – soils, evaporites, coals, glacial deposits and so on. In a 1956 contribution he concluded that the magnetic field had been maintained as a predominantly axial dipole for many hundreds of millions of years, even as far back as that contained in the billion-year-old Assynt Torridonian sandstones that he had studied for his Ph.D. a few years earlier. In his own words this conclusion was accompanied by an undeniable fact (my emphasis in italics):

> The position of the magnetic pole calculated on this axial dipole assumption, from the magnetization of Pre-Miocene age [*c.* 20 million years ago], differ greatly from the present geographic pole, polar wandering having occurred. *The poles given by data from four continents do not however agree with one another.* The discordance between the poles calculated from the magnetic data from Palaeozoic rocks of North America and Europe is particularly

notable, since it is shown by several determinations made by independent observers. *Agreement could be achieved by supposing that North America was closer to Europe in the Palaeozoic and early Mesozoic.*

This understated last sentence (see Figs 6.3, 6.4) was, along with Runcorn and Creer's work, a key factor in the (eventual) downfall of all the critics of Wegener's great achievement – continental drift *had* occurred.

One imagines both Alfred Wegener's and William Gilbert's ghosts chuckling backstage.

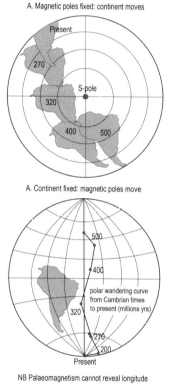

Figure 6.3 To illustrate that, in the absence of critical tests (see Fig. 6.4), palaeomagnetic data for changing magnetic vectors to the poles cannot distinguish between magnetic polar wandering and continental drift. A. Assumes fixed magnetic poles and moving continents. B. Assumes fixed continent and moving magnetic poles, so defining a polar wandering path. Redrawn from Creer (1965) and Keary et al. (2009).

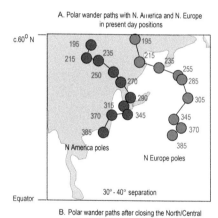

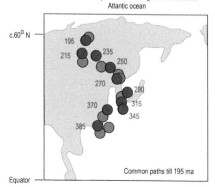

Figure 6.4 The proof of the palaeomagnetic pudding. A test to determine that continental drift had occurred can be obtained by comparing apparent polar wander paths across the Atlantic between North America and northern Europe. The discrepancy of the paths between the two land masses disappears once the modern Atlantic is closed (simplified after McElhinny and McFadden, 2000; Keary et al., 2009).

PART 3

Hot Stuff: Settings

Arthur Holmes (1890–1965) and Norman Bowen (1887–1956)

In the decades spanning the turn of the twentieth century, ingenious experimenters made use of James Clerk Maxwell's predictions concerning the existence of electromagnetic waves by using electrical discharges to generate such fluxes – the various 'rays' as named at the time – cathode, X, alpha and beta. These could interact with elemental and mineral solids and gases in various ways that revealed their atomic make-up. The results confirmed Epicurus's brave theoretical notions of atomism and added some – discovery of the phenomena of radioactive decay and heat emissions by Marie and Pierre Curie; mineral crystal structure by father and son, William and Laurence Bragg; atomic structure by Rutherford, his collaborators and students. These, *en masse*, gave rise to the discovery of entirely new radioactive elements and their isotopes, generating the new science of atomic physics that revitalized classical physics and chemistry.

Knowledge of the physical state of the earth, especially its thermal constitution and longevity, were major achievements in the decade after Ernest Rutherford and Frederick Soddy first proposed their new view of atomic radioactive decay, which enabled huge steps in understanding the geology of our planet without recourse to scripture and superstition. Arthur Holmes was one of these pioneers for his establishment of a rigorous, stratigraphically controlled radiometric timescale for a large part of the planet's existence – arguably one of the greatest achievements of the earth sciences. As a research student of R.J. Strutt at Imperial College, London, he made precise and accurate chemical analyses of the lead content of uranium minerals and, using Rutherford and Soddy's law of radioactive decay, he was able, by a judicious choice of field samples, to publish the results in a concise book, *The Age of the Earth* in March 1913. There is quite a backstory to Holmes' breakthrough and in his careful telling of it, involving two other individuals, Bertram Boltwood and Strutt himself. He later (1928–31) made major contributions to the controversy arising from Wegener's work by proposing a role for mantle convection in driving the drift of the continents, and in the process whereby heat is released from the earth's interior.

The nature of fire itself was another attribute dealt with by Epicurus, and it was the search for an understanding of the fire-formed rocks of the igneous clan that our second pioneer of this section made the object of his lifetime's work. He was the Canadian, Norman Bowen, just three years older than Arthur Holmes, and whose fieldwork forays into the backwoods of northern Ontario led him into experimental studies of the crystallization from melt of the silicate minerals that make up such common igneous rocks as granite and basalt. The molten origins of the igneous clan were once regarded by Gottfried Werner's legions of Neptunists as products of aqueous precipitation in ancestral oceans. Yet, replacing aqueous precipitation by the intrusion and extrusion of cooling molten magma raised difficult problems of interpretation. For, unlike the precipitation of natural salts from seawater or the laying down of calcareous tufa from emergent springs in limestone country, the process of melting and cooling involved in the creation of crystalline rock was impossible to observe directly. As we shall see, experimentation with a firm basis of thermodynamics was the only way out, courtesy of Bowen and of the unique experimental facilities afforded by the Andrew Carnegie-financed Geophysical Laboratory in Washington D.C.

7

Arthur Holmes (1890–1965)
English geologist

Holmes in 1912, aged 22. Wikimedia Commons.

He established the first geological timescale from radiometric ages gained from the chemical analysis of carefully selected rock samples, mostly from the Oslo rift of southern Norway (1911, 1913). He later proposed a thermal mechanism for continental drift involving the rising and falling motion of convection cells in the earth's upper mantle (1928–31, 1944).

Who was Arthur Holmes?

In the winter of 1911, while Alfred Wegener was researching his new geological hobby in Marburg, Holmes was a young Ph.D. student in London, setting out his stall in the Royal College, London over an equally controversial issue – the age of the earth. In a seminar on January 30th to the Natural History Society of the college he revealed some of the results of his long labours over the past months in the analytical laboratories of his research supervisor, R.J. Strutt. He had just obtained the first reliable and geologically meaningful radiometric ages by utilizing chemical measurements of uranium-bearing minerals that recorded the decay of unstable radioactive uranium to stable lead. The minerals were separated from judiciously chosen rock samples and enabled the drawing up of the first of successive and authoritative geological time series. Together with his other geological accomplishments – the origins and nature of igneous rocks and the role of convection currents driving continental drift – it established him as one of the greatest of twentieth-century earth scientists. Like his near-contemporary, Reginald Daly of Harvard, whilst at the universities of Durham and Edinburgh, he also popularized geology through the writing of two editions of *Principles of Physical Geology* (1944, 1965) – perhaps the most influential, inspiring and wide-ranging geological exposition since Charles Lyell's blockbuster, *Elements of Geology* (1838).

A year later, in 1912, when Wegener was overwintering on the Greenland ice cap and Norman Bowen (Chapter 8) had completed his first melting experiments, Holmes was busy writing the preface to his epoch-breaking book *The Age of the Earth*, which appeared in the bookshops of London and New York in March 1913 (Fig. 7.1). The preface begins in a delightful way:

> It is perhaps a little indelicate to ask of our Mother Earth her age, but Science acknowledges no shame and from time to time has boldly attempted to wrest from her a secret which is proverbially well guarded. On January 30th, 1911, I placed before the Natural History Society of the Royal College of Science a brief account of some of these attempts, and out of that paper this little book has gradually developed.

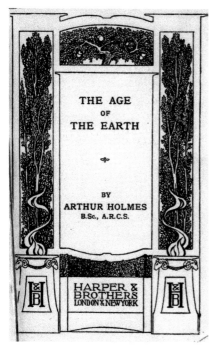

Figure 7.1 The frontispiece to Holmes' 1913 book – a suitably avant-garde production from the American publishers of what turned out to be one of the most significant scientific books ever published. Holmes was one of the few scientists to publish such a *magnum opus* at such a young age.

As we shall see, Holmes was a gifted writer for a young man of just 23, pitching his accessible English prose to both a geological audience and a general one. He presented a careful and fair assessment of the history of scientific attempts on mother earth's 'secret', also giving a clearly reasoned account of the methodology involved in his own radiometric age dating.

Holmes was born in 1890 and grew up in Gateshead, County Durham. His family were artisans on the paternal side, his father a cabinet maker and his mother a teacher. Their terraced family home in Hebburn is now the proud bearer of the blue plaque of municipal recognition familiar to UK readers. Although staunch Methodists, it must also have been a musical household, for Arthur, a star pupil in science at Gateshead High School taught by a gifted physics master, also had a lifelong love of music. He played piano in middle life well enough to charm a future wife and had great admiration for contemporary composers. For example, as his Royal Society biographer, Kingsley Dunham (himself an organ scholar and a successor to Holmes at Durham) recounts, he was a great fan of Stravinsky and is remembered in his Edinburgh years wildly applauding, standing on his seat, at the close of a concert performance of the composer's music.

In 1907, based on a National Scholarship in Physics, Holmes entered the Royal College. Like those of earlier generations (e.g., Richard Oldham), he found the scientific atmosphere at that place, (the brainchild of Thomas Huxley, Henry de la Beche and Prince Albert in the 1850s) much to his liking, graduating in geology and chemistry in 1910. He was taken on as a Ph.D. student by the shy but kindly and charismatic occupant of a newish chair in physics, Professor the Honourable R.J. Strutt whose brilliance as a radio-isotope pioneer features below – Holmes was his first Ph.D. student at the College.

He completed his intensive analytical work on the uranium and lead ratios of carefully chosen minerals from Norwegian rocks (see below), leaving his results with Strutt, when in March 1911, due to personal financial difficulties, he took up a post as geologist on a mineral exploration expedition to Mozambique. There he contracted blackwater fever (a dangerous variety of malaria) and returned to England nine months later, lucky to have survived. From 1912 to 1920 he was employed as a demonstrator (student instructor/researcher) at the Royal College and subsequently, for two years, as chief geologist to a Burmese oil company. After difficult post-war years seeking employment as an academic geologist, and the personal anguish of the early death of his young son (he had married in 1914), he was appointed to the founding Chair of Geology at the University of Durham in 1924.

Earlier estimates of earth's longevity

In the spirit of Holmes' own efforts in his *Age of the Earth* it is appropriate to take a brief review of the history of opinions on this subject as he himself had found as he began his studies at the Royal College. 1913 was a little more than 100 years or so after Playfair's version of James Hutton's *Theory of the Earth* was published. Hutton's tantalizing and much misunderstood *caveat* in his first version of 1788 refers to the geological record of our planet as revealing '...*no vestige of a beginning, – no prospect of an end.*' This magisterial dismissal of John's New Testament *Revelations*, 'I am alpha and omega; beginning and end', challenged and (eventually) triggered the death of bible-based earth history, announcing instead the birth of an observationally based science within a cyclical and analytical world view. But estimates of earth's longevity or even of the duration of his earth cycles meant little to Hutton, who, as noted previously, regarded the planet just as he would an efficient machine.

In the Jewish and Christian faiths (and wider in society at least until the Age of Enlightenment), a given was that the creation of the earth took place over a brief time interval. That and the ensuing Great Flood determined the acceptable boundaries into which early geological observations could be placed until at least the middle of the nineteenth century. The longevity of the earth from rapid creation to its drowning was a more complex matter, obtained by counting the years from the generally accepted date of Christ's nativity, 4 BC, backwards between editions of the Torah, the dates of the Romans, the destruction of the Temple and of the generations whose recorded lives and descendants could be agglomerated into a convenient number. Many divines and lay persons had a go at the calculation, including Renaissance scholars like Julius Scaliger and Isaac Newton, the latter privately as part of his enquiries into the origins of 'true' (Unitarian) Christianity. Bishop Ussher's famous estimate of 1650 was described by him as: 'the entrance of the night preceding the 23rd day of October... the year before Christ 4004'.

Ussher and others' investigations did not lead them to the longest and greatest archive of past human time, that of nilometers. These were gauges within permanent sumps in the river, used to record the water level at the high point of its annual flood. Their purpose was for tax-gathering, and they operated for several thousands of years in both Pharaonic and Ptolemaic Egypt, providing information with which the ruling classes could judge the degree of agricultural productivity each year – the ensuing taxable payment was directly proportional to the Nile flood sediment discharge and therefore the magnitude of irrigable water and nutritious sediment deposition.

In modern times a natural record of annual events, a sort of glaciometer, came along after the discovery of ice ages, when Swedish glacial lake deposits were quarried for aggregate in the 1880s. Their stratigraphy revealed to Gerard de Geer the record of successive spring meltwater deposits of coarse sediment layers known as varves – which by judicious counting eventually led back for several thousand years. As geological exploration continued into the twentieth century and became more precise, stratigraphic sedimentary rock successions many kilometres thick paid testimony to the incomparably long time intervals needed for their deposition compared with both the records of nilometers and of varves. Values of mean denudation of the continents were also calculated from sediment and solute yields of the world's major rivers, giving values for mean denudation of around a metre every 25 000 years. This would mean that 5 km of mountainous relief would, on average, vanish every 125 million years. Such estimates attested to the probability of a very large longevity for the earth but could not be used to obtain a firm estimate of just how long. Similar strictures applied to the use of sedimentation rates gathered over the whole of recognized geological time.

William Thomson's computations

A crisis of geological ideas versus religious belief was eventually reached in the mid-nineteenth century when two promising avenues for enlightenment opened. The first came courtesy of the Irish-born physicist, William Thomson, one of the originators of thermodynamics. He held a good deal of animus towards geologists for their lack of competence in mathematical physics, strongly disagreeing with the core of Charles Lyell's geological philosophy – uniformitarianism. This demanded unchanging rates of natural processes and conditions over time. Thomson saw in its uniformity a naivety of basic physics that contradicted the fundamental laws that he himself had laid down in the 1850s, chiefly the Second Law of Thermodynamics, which defined the increasing disorder (*aka* entropy/thermal decay) of natural matter. He wrote on the law's challenging implications for the longevity of the Earth and of Lyell's geology:

> Within a finite period of time past the earth must have been, and within a finite period of time to come the earth must again be, unfit for the habitation of man as at present constituted, unless operations have been, or are to be performed which are impossible under the laws to which the known operations going on at the present in the material world are subject.

In a paper published in the *Philosophical Magazine* for 1863, 'On the secular cooling of the earth', Thomson worked through a theory whose physical elegance contained two major assumptions. He reasoned that since the earth was once molten (assumption 1) it would solidify and cool down by simple conduction (assumption 2) and so he applied Joseph Fourier's law of conductional cooling to the problem. This stated that the flow of heat from any hot object was proportional to its internal temperature, the gradient of this temperature outwards to a cooling surface (earth's surface in this case) and the thermal conductivity of the cooling solid in question. My grammar school physics master compared this to calculating the changing surface temperature of a newly boiled egg as it lies snuggly in its cup.

Little was then known about the effects of pressure on both the ambient temperature in the central earth, its outward decrease and the average conductivity of crustal rock. Despite these uncertainties and in the best tradition of 'order of magnitude' physics, Thomson proceeded with his calculations by assuming that these quantities were as measured in deep mines and in the laboratory. He came up with an estimate of longevity of around 100 million years but with probable uncertainties that ranged well over an order of magnitude (20–400 million years). Later in life he revised and refined his calculations downwards, though none of his estimates were ever enough to satisfy the more perceptive geologists. Thomson took particular issue with Charles Darwin's geological and biological theories as advanced in his *Origin of Species* of 1859, finding objectionable both the unlimited time demanded and, perhaps more important to this Presbyterian Ulster Scot, the omission of any directional (i.e. divine) influence.

Thomson lived long enough, ennobled as Lord Kelvin in 1892, to witness both the development of convection as a potent alternative method of heat redistribution to conduction, chiefly as devised by R.S. Strutt's father, Lord Rayleigh, and to hear Ernest Rutherford lecturing on his recent research on radioactive radium's immense heat production. The latter had profound implications for earth's longevity – directly contrary to the chief assumptions of Kelvin's conductional cooling theory. It is well worth repeating Rutherford's amusing recollection of a lecture he gave at the Royal Institution in 1904 during a visit from his then workplace at McGill University, Canada – Kelvin sitting prominently in the steeply banked auditorium:

> I came into the room, which was half dark, and presently spotted Lord Kelvin in the audience and realized that I was in for trouble at the last part of my speech dealing with the age of the earth, where my views conflicted with his. To my relief, Kelvin fell fast asleep, but as I came to the important point, I saw the old bird sit up, open an eye and cock a baleful glance at me! Then a sudden inspiration came, and I said Lord Kelvin had limited the age of the earth, *provided no new source* [of energy] *was discovered*. That prophetic utterance refers to what we are now considering tonight, radium! Behold the old boy beamed upon me. (see Eve, 1939, p.170)

Salts of the earth: Edmund Halley and John Joly

At the very dawn of both the twentieth century and the age of nuclear physics, an Irish physicist and geologist, John Joly, Chair of Geology at Trinity College, Dublin came up with the idea that ocean water might hold a clue as to the longevity of the earth. He reasoned in 1899 that its present salt content must have built up over time, and from a knowledge of the amount of salts in the annual discharge of the world's rivers, some simple long division would give an indication of oceanic longevity. In fact, Joly was simply re-iterating an idea put forward in principle almost 200 years earlier by the astronomer Edmund Halley. In 1715 he had written a paper entitled: *A short Account of the Cause of the Saltness of the Ocean... with a Proposal ... to discover the Age of the World.* He wrote, with amazing prescience for the time, that in the case of lakes with no outlet:

> ...the saline particles brought in by the rivers remain behind, while the fresh evaporate; and hence it is evident that the salt in the lakes will be continually augmented and the water grow salter and salter.

And that by applying the same principle to the oceans:

> It is not improbable but that the ocean itself is become salt from the same cause, and we are thereby furnished with an argument for estimating the duration of all things.

By Joly's day it was possible to quantify the two variables involved. The methodology was seemingly logical and with his estimates falling close to the mean of Kelvin's entirely independent ones they gave rise to high hopes in both men of the correctness of their approaches. Yet stratigraphic geologists were not satisfied with either.

Joly, like Kelvin, fought long and hard against the radio-activists (if we can call them that) in the cause of his methodology, this despite his great interest and correspondence with Rutherford on radiation generally and his original work on the effects of uranium decay leading to the bombardment by alpha-particles that visibly disturbed the atomic structure of minerals like the micas. His opposition lasted to his death in 1933, by which time he was better known as the inventor of a technique for colour printing and, perhaps more profoundly for society and health, as the co-founder of radiotherapy with the establishment by the Royal Dublin Society of the Irish Radium Institute, which pioneered the 'Dublin method' that gradually spread worldwide.

With hindsight we can understand better why Joly's methodology had a major and fatal flaw. In common with most of his contemporaries he had envisaged the oceans as one-way receptacles – 'sinks'. We now know that they are both sinks *and* sources – two-way recyclers of elemental mass, constantly interacting with the atmosphere and with the sediment and lavas on the floors of the ocean to arrive at a chemical composition that is in a state of dynamic equilibrium. Although these are modern ideas, the notion of recycling oceans was not a new one, and it was to be tentatively explored by Holmes in *The Age of the Earth*. Even so, Holmes was a deadly reviewer and closed his review of Joly's approach in typically visionary style:

> The whole discussion merely serves to betray the uncertainty of the method and the doubtful applicability of even the most accurate data. For the present we can only conclude that our knowledge of the part played by sodium and chlorine in the constant redistribution of the materials of the earth's crust is still lamentably imperfect, and that quantitative deductions drawn from it must be regarded as being purely provisional.

Joly's values, even with the cyclical aspects explored by Holmes, were gross underestimates of the planet's longevity, adding to the angst of geologists and palaeontologists in the age of Darwin when vast amounts of time were deemed necessary for the evolutionary process inherent in natural selection to run its course. However, help was at hand.

Radio-activists: natural minerals and unstable elements

When I was a sixth-form student, fired up by the subjects of geology and chemistry, I became a fanatical mineral collector. A summer trip to mineral-rich Cornwall, youth hostelling with my brother in the summer of 1965, enabled me to collect samples of uranium-bearing minerals from the famous South Crofty mine's spoil tips. Back at school in September my physics master kindly produced his Geiger counter, whose rattlesnake-like reaction to my samples confirmed both my identifications and their radioactive properties. I mention this slight personal experience because not only had Marie and Pierre Curie obtained supplies of uranium ore from South Crofty, but also to the fact that the element and its various natural compounds had long been known to chemical and mineralogical science. Its various laboratory-produced compounds owed their diversity to the many oxidation states of the uranium ion – chlorides, sulphates, carbonates, nitrates and halides. As is well known, the discovery of radioactivity was a curious amalgamation of serendipity in laboratories that involved cathode ray tubes, photographic plates and artificial uranium salts. These led to the discovery of evanescent X-rays and alpha rays (eventually identified by Rutherford as helium ions). These discoveries kick-started the whole radioactive revolution, spawning some notable revolutionaries, including two prominent 'radio-activists', one American and one English, who were at the centre of efforts in applying its basic tenets to the rocks of ages.

B.B. Boltwood's analytical genius

Bertram Borden Boltwood, Amherst-born and a Yale professor for most of his career, was a dedicated experimentalist of independent means who fitted out and maintained his own laboratory as a young consultant. His exact mastery of the black arts of 'wet chemistry', his facility in improving practical apparatus and his thorough knowledge of mineralogy enabled him to contribute massively to the early development of ideas concerning the age of the earth as revealed both by the decay of uranium to radium and helium and by its eventual decay to stable lead. He became interested in developments in the radioactive field from around 1900 that led naturally on from his previous studies of rare-earth elements. Whilst investigating radioactive substances in pitchblende (uranium oxide, the chief mineral ore of uranium) he discovered the 'element' ionium, later determined as an isotope of the element thorium, symbol ^{230}Th. Here is Boltwood's obituarist (Kovarik, 1929) commenting on this early part of his career:

> Radioactivity at that time was not a science as yet, but merely represented a collection of new facts which showed only little connection with each other. These facts needed some correlation and explanation with some basic hypothesis before they could give, as a whole, a semblance of a science. The initial placing of radioactivity on a scientific basis came with the

announcement by Rutherford and Soddy in 1903 of their theory of disintegration of radioactive elements. Briefly stated, this theory postulates that an atom of a radioactive element, e.g., uranium or radium, spontaneously disintegrates emitting energy in the form of radiations and that from what is left of the initial atom, an atom of a new element is formed which may in turn disintegrate. This was a bold theory in 1903 but today it is no mere theory, but an established fact verified in every case. To this verification Boltwood contributed early and materially. He devoted much time to the investigation of problems dealing with the origin of radioactive elements and with the genetic relationships among these elements.

Boltwood's breakthroughs into radioactive research came by 1904, after he had made an intensive study of the ratio between uranium and radium in naturally occurring minerals. Again, his obituarist notes:

> In his papers [at this time, 1904–1907, some jointly with Rutherford] he demonstrated that radium must be a disintegration product of uranium. He had analysed primary uranium-bearing minerals from all parts of the world and showed that the ratio of the activities of radium and uranium was remarkably constant for all old unaltered minerals from various geological formations and with widely varying amounts of uranium content.

There was simply no other explanation for the constancy of that ratio in old rocks, given the proposal by Rutherford and Soddy. Thus was born the concept of a 'decay series' as it was later called. It turned out to be a constant ratio for rocks of a given age, a fact shared by the relationship between uranium and lead and whose fundamental consequences for geochronology were to be noted by Boltwood and thoroughly explored by Holmes for his Ph.D. thesis. As Boltwood wrote in his 1907 paper:

> Out of a considerable number of analyses undertaken with the particular object of discovering whether or not lead was present, I have been unable to find a single specimen of a primary mineral containing over 2 per cent of uranium in which the presence of lead could not be demonstrated by the ordinary analytical methods...

> ...Through a dawning appreciation of the significance of the persistent appearance of this element in uranium minerals, the writer was led to suggest in an earlier paper that lead might prove to be one of the final inactive disintegration products of uranium. All the data which have been obtained since that time point to the same conclusion.

Brilliant loner: The Honourable Robert John Strutt (*aka* the 4th Lord Rayleigh)

Strutt was born into the British scientific aristocracy as a real aristocrat, the eldest son of the 3rd Baron Rayleigh, a hereditary peer and Nobel prizewinner for his co-discovery of argon. Rayleigh Snr had broad interests and was renowned in physics circles, with his name appended to such phenomena as Rayleigh scattering (why the sky is blue), Rayleigh surface earthquake waves, the Rayleigh dimensionless number (whether a substance will convect or not) and of Rayleigh-Taylor flow instabilities (rollers and waves). The eminent physicist, J.J. Thomson, was his pupil and the future mentor and supervisor of Strutt, who eventually (in 1919) succeeded his father to the baronetcy, a quaint illustration of the easy British caste system. He then gave up his chair and labs at the Royal College and conducted various lines of research from a well-equipped home laboratory originally constructed by his father. Strutt's scientific *oeuvres* come under both names, Strutt and Rayleigh, the former dominating his early career with its noteworthy contributions to early atomic and radioactive research, and later, in the application that concerns us here, his thorough examination of the application of helium-dating to rocks. Also, as we have mentioned, most relevant to our main theme, he was Arthur Holmes' Ph.D. supervisor.

In work from 1899–1903 at the Cavendish Laboratories in Cambridge on the effect of Becquerel rays (Rutherford's beta rays) on various gases under magnetic fields, Strutt had attributed easily deflected rays from radium to these negatively charged atomic items (later known as electrons). With Rutherford he demonstrated (in the presence of a delighted Kelvin), using a gold leaf electroscope, that these signified the immense amount of energy present in the element. He also became fascinated at this time by the occurrence of helium and radium in the natural spring waters of Bath (Fig. 7.2) and widened the scope of his interest into the occurrence of the gas in rocks and minerals – helium was known naturally up to then only from solar spectral emanations. Closely following similar determinations by Boltwood, the constancy of the radium/helium ratio became an established fact and led to the supposition by Rutherford that the gas was a disintegration product of the metal.

Strutt's first real coup in the fast-moving field of radioactivity as applied to the earth was a brilliant 1906 assessment of the contribution of radium decay to the global heat budget. In the four years previously Kelvin's heat-loss model for the earth had suffered several mortal wounds from young radio-activists like Strutt. He now proceeded to give it the *coup de grace* from his detailed chemical analyses of various igneous rocks and some telling calculations. He argued that their abundant radium content, greatest in granitic examples, and the assumption of their sourcing by melting of the topmost mantle and crust were entirely sufficient to maintain the earth's temperature. His final calculations demanded that *all* the planet's radium-induced radioactivity and heat was generated in a thinnish 100 km or so outer layer, his crust. Later in life, Strutt/Rayleigh remarked that he was inclined to think that the result to which he had been led was perhaps the most important of his contributions to science.

Figure 7.2 The famous waters of Bath Spa, Somerset, England in their Georgian-framed Roman setting. The 3rd Lord Rayleigh 'took the waters' here and in them his son, R.J. Strutt, detected the presence of radium and helium.

Strutt's full attention was now taken up with his idea, made independently by Rutherford, that helium decay from radium could be used to determine the age of rock-forming minerals. He later wrote in his clear English:

> The helium method of determining age is most strictly applied by taking an arbitrary quantity of the mineral, measuring the volume of helium, V, initially present, and the volume, v, produced in one year. Supposing the gas to have been all retained, the time required to accumulate it would be $V = V/v$ years. If the gas had not all been retained in the mineral, then the time will be greater than this, to what extent it is impossible to say [see below]. The method only professes to give an inferior limit to the age. However, subject to this limitation, it merely requires the comparison of two volumes, and is then quite independent of any detailed theory of radio-active processes. At the time when these processes were not so well explored as now, this was a valuable feature, and I think I may say that the work did something to convince the scientific world generally that there was not much that was dubious or hypothetical about the minimum estimate obtained. It showed at least that the age was much longer than Kelvin had supposed.

He had found after much experimentation with many radium-bearing minerals that gas leakage had almost certainly occurred – as inferred from the presence of inherent mineral features like post-crystallization chemical changes along fractures. He nevertheless found ages of several hundred million years, far in excess of those given by Joly's and Kelvin's estimates. His best, and significantly older, results were from chemically resistant zircons (zirconium oxide) obtained from Precambrian granites (minimum ages of 600 million years) and, from geologically older Archean examples, using the mineral sphene (minimum ages of 700 million years).

That was nearly all from Strutt concerning the age of the earth's rocks. Fully realizing the limitations of his methodology, Strutt nevertheless recognized the greater accuracy of estimates of age obtained by determining their lead content *à la* Boltwood. Shortly after his move to take up his new chair at the Royal College, after a further paper (1910) on the general subject, in the words of his obituarist 'the discovery of "active nitrogen" led him to set it all aside...for the main of the work had been done.'

Arthur Holmes: elementary my dear Strutt

What happened next was extraordinary. The young, tenacious, and analytical chemistry-trained Arthur Holmes, with a matching devotion to geology, had just completed his bachelor's degree at the College. He took the uranium–lead route recommended by Rutherford to Boltwood in 1907, despite the problems remaining associated with uncertainties regarding the contribution of thorium to what was slowly becoming a 'spent lead bank' and to eventually apply it to carefully selected rocks from different parts of the geological column across the world. This was Holmes' contribution – his chemical foresight and geological knowhow coming together to attack the core of the problem. Boltwood, Strutt, Rutherford and Soddy had been content in proving some much older ages for the earth than traditionally believed – they were no longer interested in 'details', as they saw it and moved on to more mainstream and exciting adventures and their real interests – the increasingly dark realms of subatomic physics and chemistry.

After many months of hard grind in the analytical labs of the Royal College, doing tedious but necessary conventional 'wet-chemistry' analyses, Holmes' first scientific paper was communicated by Strutt to the Royal Society in March 1911, read by the author to the Society that April. It seems likely that the aforementioned talk that Holmes gave to the Natural History Society of the Royal College would have been based on this paper, perhaps accompanied by some of the comments in a subsequent note to the *Nature* journal of a more general kind. As noted in his acknowledgements to the 1911 paper, the subject of Holmes' research was not self-generated but suggested by Strutt – a re-examination of the 1907 results of Boltwood based on theoretical and experimental rates of helium and lead production from uranium given by both him and Rutherford. The research was necessary according to Holmes because:

> for minerals of the same age, the ratio Pb/U should be constant...Further, for minerals of different ages, the value of Pb/U should be greater or less in direct proportion to those ages. Collecting all the known analyses of primary uranium-bearing minerals which included a determination of lead, Boltwood showed that the above conditions were generally found to hold. Unfortunately, he omitted to give the geological ages of the several occurrences. In a summary of his analyses, to be given in a later section, these will be indicated as accurately as at present is possible.

Holmes carefully planned an attack on the problem. He separated uraniferous minerals from igneous rocks (Fig. 7.3) collected by pioneering Norwegian geologists, Waldemar

Brögger and Arvid Högbom from the Oslo graben of Norway. The relationships of these with ancient Precambrian rocks (older than *c*.550 million years) and younger rocks and strata well-dated by Palaeozoic age (pre-250-million-year-old) fossils) were made absolutely clear by geological mapping and stratigraphy. Here is Holmes' geological rationale for the whole project:

> From these considerations, it is obvious that the only minerals to be chosen are fresh, stable, primary rock-minerals. Secondary and metamorphic minerals could not be relied upon to satisfy the required conditions. There occurs in the Christiania [Oslo] district of Norway, a geologically depressed area of nearly 4000 square miles [the Oslo rift valley], which is separated on every side by faults from the surrounding Pre-Cambrian gneiss. In this area there is a nearly complete sequence of early Palaeozoic rocks. Above these strata there are a few beds of red sandstone of Lower Devonian age. Over these beds and intercalated with them are lava flows; and, finally, penetrating the whole mass, representing a later phase of this period of igneous activity, are great intrusions of plutonic rocks. Amongst the earliest of the intrusions is a series of thorite-bearing [a radioactive mineral] nepheline-syenites. Brögger believes them to be of Middle or Lower Devonian age, most probably the latter. The minerals occurring in them are, in many instances, notably radioactive, and thus they afford an admirable series in which to investigate the consanguinity of lead and uranium.

None of the pioneer physics and chemistry radio-activists possessed the necessary geological background to have devised such a thorough testing from such promising

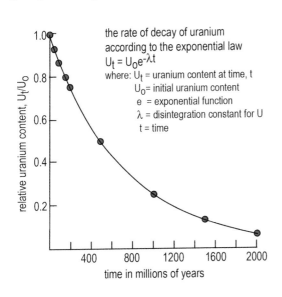

the rate of decay of uranium according to the exponential law
$$U_t = U_0 e^{-\lambda t}$$
where: U_t = uranium content at time, t
U_0 = initial uranium content
e = exponential function
λ = disintegration constant for U
t = time

Figure 7.3 Version of Holmes' 1913 illustration of the general principles for the exponential nature of radioactive decay laid down by Ernest Rutherford and his students some five years earlier. As discussed in the text, Holmes was able to derive ages of stratigraphically defined Palaeozoic rocks from the Oslo rift and of older rocks from elsewhere – the basis for the first geological timescale. Previous ages were obtained from randomly selected samples with no stratigraphic control.

Figure 7.4 The Palaeozoic and early Mesozoic igneous rocks of the Oslo rift include two iconic rock types familiar as decorative stone. A. Rhomb porphyry, showing exquisite large feldspar crystals in a finely crystalline groundmass. Sample is 6.5 cm long. B. Larvikite, named from a plutonic mass quarried at its *locus typicus*, Larvik, in southwestern Vestfold County, southern Norway, dates to about 292–298 million years old (earliest Permian). This polished sample shows its characteristic shimmering feldspar crystals up to 2 cm length with smaller black pyroxenes scattered around. Wikimedia Commons. Author: James St John.

rocks in such a closely studied and mapped geological terrane. The value of such rocks was recognized by Charles Lyell, who visited the Oslo graben in 1837 and who furnished descriptions of the geology in his 1838 edition of *Elements of Geology*, a book avidly read by the young Charles Darwin on his return from the *Beagle* years, and, perhaps, years later during his undergraduate career, by Holmes himself.

Arthur's first dates

These were as follows, with the lead/uranium ratio and age in millions of years (bold) in brackets after each geological interval sampled:

> *Carboniferous of Oslo graben* (0.041, **340**); *Devonian* ditto (0.045, **370**); *Pre-Carboniferous* ditto (0.050, **410**); *Silurian or Ordovician* ditto (0.053, **430**); *Pre-Cambrian of Sweden* (0.125/0.155, **1025/1270**); *ditto United States* (0.160/0.175, **1310/1435**); *ditto Ceylon* (0.20, **1640**).

So, 100 years or so after Playfair rewrote the Huttonian message of what we now call 'deep time', and three years before World War I, humankind learnt that the planetary crust was unimaginably older than ever realized.

Holmes' Devonian result, though not the oldest date by any means, was perhaps the most pleasing and noteworthy. The mineral sample came from part of an igneous intrusion into fossiliferous sedimentary strata that contained Lower Devonian fossil fish. The igneous intrusion therefore had to be younger than these fossils and their entombing strata – it was dated by Holmes to 370 million years ago (Ma). This is within the span nowadays recognized as the time limits of the Devonian period, 420–360 Ma, appropriately in its uppermost part.

Holmes' pioneering result was wonderfully accurate and a tribute to his painstaking analytical techniques. It truly helped to change our world view – not a bad outcome for a young research student.

Holmes' earliest radiometric timescale with its tale of hundreds and thousands of millions of years of earth history bewildered and upset many, not just lay persons and clerics, but also geologists of the 'old school' and, perhaps, the ghost of Lord Kelvin. Here is the barbed response to such sceptics by the well-known palaeontologist and author Sir Arthur Trueman in the 1930s, writing in the preface to his ever-popular *Geology and Scenery of England and Wales*:

> And if there are readers who would say that scientists have no right to talk in terms of those inconceivably long periods we may point out that the geologist is as well able to appreciate the hundreds of millions of years representing the age of the earth as the Chancellor of the Exchequer [the archaically-named office of British Finance Minister] is able to appreciate the total of his budget.

Geologists now have four and a half billion years to play with, according to the latest radiometric determinations of lead in ancient zircons. Modern Chancellors in these post-COVID times now talk in terms of tens of billions – appropriate for the 14-billion-year age of the universe.

Holmes on convection, drift and heat loss

Picking up Holmes' research agenda after his 1924 appointment to the Durham chair we see a man determined to pursue the implications of radioactivity to wider geological processes. By this time he also possessed an added sense of geological grandeur, having worked in the vast and ancient igneous and metamorphic landscapes of SE Africa and from which he had collected innumerable specimens for analysis. At the Royal College he gradually became an expert in the field of igneous rock studies, publishing definitive books on the subject in the early 1920s. There was one extra to add. He now had the time to explore the consequences of Wegener's drift theory, a topic first introduced to him during the war years by his polymath colleague, John Evans, who had done much in the way of helping J.G.A. Skerl in his translation of Wegener's third edition of 1924, writing the introduction and corresponding with Wegener concerning it. It was the first English version of Wegener's *opus* published two years earlier and, as we saw in Chapter 5, it aroused much interest and controversy in the post-war geological world of the Anglosphere.

By 1928 Holmes had put together perhaps the most influential and authoritative

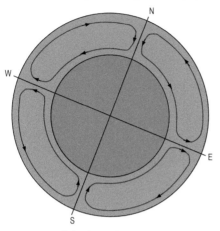

A. General idealised planetary circulation by mantle convection: descending currents at poles; ascending currents at equator

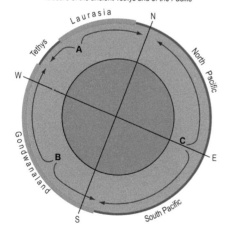

B. Actual convective circulation: ascending vigourous currents under hotter [more radioactive] continental crust at A & B cause continental separation of Gondwanaland and Laurasia, closure of the ancient Tethys and of the Pacific

Figure 7.5 A. In the late 1920s, Holmes envisaged whole-mantle convection with a pole-to-equator positive gradient in mantle heat production – his elongate convection cells extend over each quadrant of the mantle. B. Holmes recognized that the distribution of the continents would play an important role in the convectional regime since the greater radioactive heat output from their crust would increase convective heat transfer from below. Here he posits convection cells amplified under crustal agglomerations such as Wegener's late-Palaeozoic supercontinent Pangaea and the Caledonian–Appalachian Mountain belt.

commentary of all thus far on Wegenerian theory. He read it as a paper to a meeting of the Geological Society of Glasgow in January of that year, three years before it could be published in that society's Proceedings. He astutely concentrated on proposing a new mechanism for the process of continental drift – *why* should continents move laterally? Briefly and authoritatively he brushed aside previous proposals based on cosmic, earth rotational, isostatic and thermal earth contraction/expansion forces as either faulty in principle or inadequate in magnitude for the purpose. Instead, he focused on the probable role of convection currents in the earth's mantle (Fig. 7.5), the general phenomenon of convection in fluids having been investigated by R.J. Strutt's father decades before. Rayleigh had assembled the variables at play in the forces that might allow convection to occur in fluids – the ratio of these forces is now known as the Rayleigh Number.

Based on chemical analyses of the common igneous rock types making up the crust and upper mantle, Holmes made estimates concerning the magnitude of radioactive heat fed through the mantle and lower crust to the outer crust (as had Boltwood and Strutt, though less exactly, 20 years earlier). He proposed four step-by-step deductions concerning heat production and the *necessity* of the process of continental drift. Slightly paraphrased and simplified they were as follows:

a) if earth's crust enables the loss of radioactive heat to the surface by conduction, and

b) the mantle has only a tiny fraction (*c*.0.001) of the heat-generating capacity of volcanic crustal basalt; then two outcomes are certain:

c) the mantle cannot yet have cooled sufficiently to have crystallized, but must still be circulating convectively; and

d) to avoid permanent heating up, some process such as continental drift is necessary to make possible the discharge of heat.

By 1944 and the publication of the first edition of his immensely popular *Principles of Physical Geology* (note the echo from Lyell's own *Principles* in his title), much of it written during early wartime fire-watching on the roof of Durham University's science labs, we see that Holmes had completely changed the mobilist research agenda in a most profound way. Continental drift was now seen as both integral to wider geology (a view shared by his friend Reginald Daly at Harvard) with the convectional cycling of the earth a *necessary* outlet (source) for radioactive heat loss. The outlet was to work by the opening up of new oceans by mid-ocean ridge crustal magma on the spreading tops of convection cells (Fig. 7.6). At the same time, to conserve earth's volume, oceanic crustal material is lost by its sinking around the peripheries of newly split continental fragments by descending and cooling parts of the convecting cells, much-aided by the deep conversion of basalt to the denser rock eclogite under high pressures. The result was the splitting apart of large continental agglomerations like supercontinental Pangaea formed 100 million years earlier, the closure of Tethys and the formation of the Alpine–Himalayan Mountain chain (Fig. 7.7).

Happy endings

Such proposals put Holmes back at the very forefront of global tectonic research for the next 30 years or so. His radiometric-dating research flourished too, despite the strains of running and setting up the new Durham department. He had also become a contented man in his personal life, for after the death of his first wife in 1938 he was able to marry the woman whom he had met and fallen in love with in the early 1930s

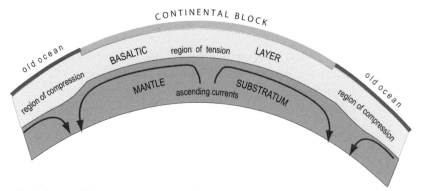

A. Generation of a new ocean: onset of convection

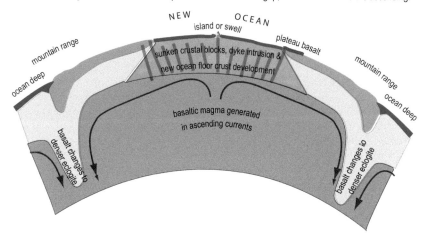

B. Continued and vigourous convection: 'with...mountain building...where the currents are descending, and ocean floor development on the site of the gap, where the currents are ascending.'

Figure 7.6 The reconstructions in Figure 5.1 and 7.5B encouraged Holmes to speculate on the likely sequence of events that might occur as a supercontinent such as Pangaea was subjected to intense convection. The early opinions in his 1931 paper were amplified in the 1944 *Principles of Physical Geology* where he presented the 'speculative' idea that new ocean crust is generated by what we now call sea-floor spreading. He shows feeder dykes to the mid-ocean ridge and 'plateau' basalt eruptions on the new ocean margins (as in the Hebrides, Greenland and Deccan, for example). To compensate for these new additions, older oceanic crust must be got rid of at an equal rate at ocean trench margins by what came to be known as subduction.

on a field excursion to Ardnamurchan. This was Doris Reynolds, a forceful and intense igneous petrologist who in the 1940s and 50s became one of the leading proponents of granitization – a non-magmatic theory for the origin of granite, an ultimately doomed concept applicable on small scales but much ridiculed in the wider scheme of things

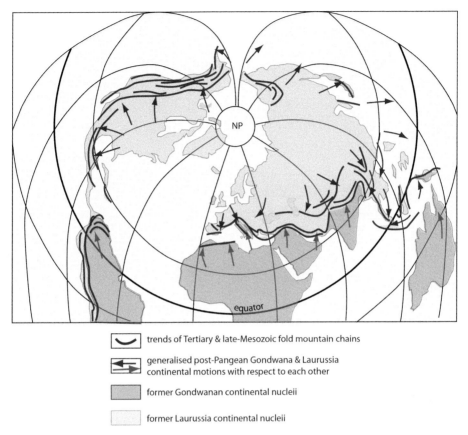

trends of Tertiary & late-Mesozoic fold mountain chains

generalised post-Pangean Gondwana & Laurussia continental motions with respect to each other

former Gondwanan continental nucleii

former Laurussia continental nucleii

Figure 7.7 Holmes' 1931 global take on Wegenerian 'drift' and the formation of the Alpine/ Himalayan, Western Cordilleran and Andean ranges because of the fragmentation and partial re-assembly (via Eurasia) of the Pangaea supercontinent.

by, amongst many others, Norman Bowen (whom we shall meet in the next chapter). In 1943 the pair left Durham for a restart at the University of Edinburgh.

By the early 1960s, in retirement after a glittering and productive late career in the now-revitalized Edinburgh department, Holmes was able to finish and deliver his fat and heavy (2.72 kg) second edition of *Principles*, published in 1965, the year of his death. This book became *the* intellectual springboard for many of the new generation who were beginning to study geology at that time and since, offering a chance to delve into the fascinating story of mobilist tectonics from one of its greatest exponents. I was one such student, and as secretary of the Durham student geological society in 1968 had the honour to write to Doris Reynolds on behalf of our committee asking for her blessing if we renamed our society the 'Arthur Holmes Society' – she gave it graciously and so it remains today.

8

Norman Levi Bowen (1887–1956)
Canadian/United States geologist/geochemist

Norman Bowen in his early days at the Geophysical Laboratory, Washington. Copyrighted image courtesy of the Carnegie Institution for Science, Washington D.C., USA.

His meticulous early experiments revealed the evolution of mineral phases in crystallizing laboratory melts designed to carefully approximate the composition of the commoner igneous rocks, so explaining some of their observed diversity (1915, 1935). In his *magnum opus* of 1928, *The Evolution of Igneous Rocks*, he saw igneous diversity (differentiation) as a consequence of what was known as crystal fractionation by the predictable separation of mineral constituents during crystallization (1928).

On things igneous

The igneous rock clan is mostly composed of minerals that have crystallized out from cooling silicate melt – magma. It might cool slowly, remaining in the subsurface as coarsely crystalline masses (plutons; cf., Pluto of the Roman underworld) or cool more quickly as finely crystalline lavas erupted from volcanic centres (cf., Vulcan, the Roman god of fire). The study of such rocks is interesting in itself, but also fundamental to our knowledge of the inner earth processes that caused the magmatism in the first place. It is also a complex subject, beset at the beginning of the twentieth century by an over-abundance of names and jargon as geological exploration of far-flung places on land and in the oceans harvested a rich and seemingly endlessly diverse haul of specimens to the world's museums and geological institutes.

But what did all these carefully collected specimens signify? Why were oceanic islands and the exposed tips of the mid-ocean ridges composed mostly of silica-poor (aka 'basic') rocks like dark-coloured basalt, whereas the continents seemed largely dominated by silica-rich (aka 'acidic') light-coloured granitic rocks? What could explain the 'order of crystallization' deduced for the common igneous rock-forming minerals from microscopic studies? Why did the melting points of mineral constituents differ so much?

As we shall see, long ago James Hall showed how igneous rocks, but also sedimentary limestones, could be melted and re-fused by slow cooling or rapid quenching. A century later, Norman Bowen, the subject of the present chapter, tried to cut through all the verbiage to explain by careful experiment the origins of the commonest igneous rocks. Bowen showed how crystallization from experimental melts produced from carefully weighed mixes of selected mineral constituents could explain some of the natural variation in mineral composition of the commoner igneous rock associations.

An early investigative tack came in the mid-nineteenth century courtesy of Henry Clifton Sorby's use of a bottom-lit microscope fitted with Nicol's prisms, which polarized incident light. This enabled examination of ultra-thin (c.30 microns/0.03 mm) sections of sliced rock whose constituent minerals could then be clearly seen on the microscope stage and their interrelations established. The optical properties of individual minerals – their colours, refractive indices and morphology – could be easily studied in the laboratory (Fig. 8.1) and, in conjunction with chemical analyses, their compositions determined. Classification schemes were set up involving the relative abundances of light-coloured minerals like quartz and feldspar compared to darker, more iron- and magnesium-rich olivine and pyroxene. A sort of crystal stratigraphy was also set up that denoted the priority of crystallization of mineral phases: intergrowths denoted coupled or tripled crystallization; the envelopment of well-formed crystals by younger, less well-developed forms; the existence of large, often perfectly formed crystals set in a younger, finely crystalline groundmass – the famous porphyritic texture of Figures 8.1B and C denoting crystallization first by very slow then by very rapid cooling.

The constituent minerals in igneous rocks have definite crystal structures of various kinds, based around the presence of a fundamental unit of one atom of the element silicon in combination with four oxygen atoms, forming four-cornered aggregates known as silicate tetrahedra. Such pyramidal forms are free to bond with other tetrahedra and with the remaining most abundant elements present in magma – including iron, aluminium, magnesium, calcium and potassium. The resulting crystalline forms have characteristic structures that were first revealed to William and Lawrence Bragg by X-rays in 1912. Subsequent decades of research established that

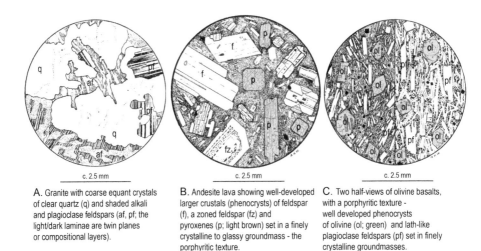

A. Granite with coarse equant crystals of clear quartz (q) and shaded alkali and plagioclase feldspars (af, pf; the light/dark laminae are twin planes or compositional layers).

B. Andesite lava showing well-developed larger crustals (phenocrysts) of feldspar (f), a zoned feldspar (fz) and pyroxenes (p; light brown) set in a finely crystalline to glassy groundmass - the porphyritic texture.

C. Two half-views of olivine basalts, with a porphyritic texture - well developed phenocrysts of olivine (ol; green) and lath-like plagioclase feldspars (pf) set in finely crystalline groundmasses.

Figure 8.1 The three commonest igneous rocks sketched from thin sections under the microscope when viewed under plane-polarized light. The texture, composition, crystal size, origin and order of crystallization of the silicate minerals involved were the subject of Norman Bowen's life work. Original sketches by A.K. Wells (a relative of H.G.); redrawn with the coloured minerals emphasized appropriately.

silica–oxygen tetrahedral (pyramidal) 'skeletons' occur in all sorts of aggregate forms: as part of random networks (the minerals quartz and olivine); in regular networks (feldspar); in sheets (mica) and chains (pyroxene, amphibole). Predicting exactly which of these kinds of minerals form, and their chemical make-up as they crystallize from cooling magma, was the core problem that Bowen set out to solve.

An experimental laird: James Hall of Dunglass

But Bowen had a noteworthy predecessor, for James Hutton attracted several younger acolytes during his life, including James Watt. Also prominent was a young Scottish laird with decided opinions – James Hall (Fig. 8.2), whose Dunglass estate was in the parish of Oldhamstocks, East Lothian, in the Scottish Borders 'Old Red Sandstone' country. This was close to the historic site of Hutton's most spectacular unconformity at Siccar Point on the Berwickshire coast (the others were on Arran and at Jedburgh). In 1788, Hall accompanied Hutton and Playfair from his estate, searching eastwards in search of just such a thing – they stumbled ashore from their small craft onto an outcrop to view the scene. Here is Playfair's rapturous recollection of it:

> We felt ourselves necessarily carried back to the time when the schistus on which we stood was yet at the bottom of the sea, and when the sandstone before us was only beginning to be deposited, in the shape of sand or mud,

from the waters of a superincumbent ocean. An epoch still more remote presented itself, when even the most ancient of these rocks, instead of standing upright in vertical beds, lay in horizontal planes at the bottom of the sea, and was not yet disturbed by that immeasurable force which has burst asunder the solid pavement [crust] of the globe.... The mind seemed to grow giddy by looking so far into the abyss of time.

The discovery must also have made a terrific impression on the 27-year-old Hall who, inspired by Hutton's grand theory (though not so much by his cumbrous prose), was a constant companion and interlocutor of Hutton in the late 1780s and 90s. Three years before the Siccar Point experience he had taken himself off on private fieldwork all over Europe (a geological 'Grand Tour'), particularly to visit active Italian volcanoes and the trans-Alpine mountains. When he was in Rome in 1785 his portrait was painted by the fashionable artist, Angelica Kauffman (Fig. 8.2). It shows the handsome 24-year-old's intense downward glance via a pair of hawk-like eyes – it is easy to imagine him 20 years later, still partly exasperated by, but also in awe of, his now-dead mentor. Looking back, he wrote:

After three years of almost daily warfare with Dr Hutton, on the subject of his theory...I was induced to reject his system entirely, and should probably have continued still to do, with the great majority of the world, but for my habits of intimacy with the author; the vivacity and perspicuity of whose conversation, formed a striking contrast to the obscurity of his writing...I thus derived from his conversation, the same advantage which the world

Figure 8.2 Angelica Kauffmann's *Sir James Hall of Dunglass, 1761–1832*. National Galleries of Scotland PG 2990. Purchased with the aid of the Art Fund 1995. Kauffman was much admired by Joshua Reynolds during her previous 15 years' stay in England. She was one of only two females to have been elected as Founder Members of the Royal Academy, London.

has lately done from the publications of Mr Playfair's *Illustrations* [John Playfair's magisterial version *Illustrations of the Huttonian Theory*, 1802]; and, experienced the same influence which is now exerted by that work, on the minds of our most eminent men of science.

Hall himself had a remarkable mind in that he had a very real sense of what was possible to achieve by experimental approaches to geological problems – a novel and ambitious approach in those days, perhaps not seen since the magnetic experiments 200 years earlier by William Gilbert. He was passionately interested in the Huttonian interpretation of the origins of igneous and metamorphic rocks. This interest obviously irked Hutton, a man who readily admitted hare-brained *ad hoc* explanations for other rock types, like a molten origin for the stratified flints so commonly seen in the undeniably sedimentary chalk deposits of Ulster and southern and eastern England. When asked by Hall in 1790 if he would support his experimental aims, Hutton declined, saying that there was enough proof existing already from field relationships and that glass furnaces could in no way duplicate those of Mother Nature. Hall's own philosophy in experimental investigation is expressed in wise words that modern experimentalists of all persuasions will immediately recognize:

> But, notwithstanding my veneration of Dr Hutton, I could not help differing with him on this occasion: For, granting that these substances, when in fusion, were acted upon by heat of ever so great intensity, it is certain, nevertheless, that many of them must have congealed in moderate temperatures, since many are fusible in our furnaces…the imitation of the natural process is an object which may be pursued with rational expectation of success; and, could we succeed in a few examples on a small scale, and with easily fusible substances, we should be entitled to extend the theory, by analogy, to such as, by their bulk, or by the refractory nature of their composition, could not be subjected to our experiments. It is thus that the astronomer, by observing the effects of gravitation on a little pendulum [He is referring here to Cavendish's experiment of 1797–98 to determine the density of the earth], is enabled to estimate the influence of that principle on the heavenly bodies, and thus to extend the range of accurate science to the extreme limits of the solar system.

Five years later Hutton wrote the following sarcastic and censorious comment on Hall's approach in Part 1 of the revised long version of his *Theory of the Earth*, when he mocks those who 'judge of the great operations of the mineral kingdom, from having kindled a fire, and looked into the bottom of a little crucible.'

It says much for Hall's admiration for Hutton that he subsequently put off carrying on with his experiments while Hutton was still alive – a mark of his respect for his master. However, the speed with which he began experimental work in 1797-8, immediately after Hutton's death, bears witness to an investigative determination that deserves our admiration down the years.

How did Hall proceed in the way of experiment? Firstly, he had given a great deal of thought to the purpose and detailed direction of his experiments, as all good experimentalists must do. In his 1790 paper, on his aborted attempt to experiment on granite melting, he had already made observations in the arrays of furnaces arranged for glass blowers in the conical glasshouses that had sprang up in the later eighteenth century in the coalfield districts of northern Britain. In these edifices coal was used to melt pure silica sands. He noted experiments carried out on melting mixtures of silica and feldspar, and in particular a molten mass accidentally left to cool that had turned into a white-coloured hard and refractory state. Hall suggested that if molten mixtures of powdered granite were allowed to cool sufficiently slowly, they might revert to their original crystalline forms of quartz and feldspar. He speculated that crystalline textures involving larger crystals set in a groundmass of smaller ones (our porphyritic texture) might record initial slow cooling at great depth followed by rapid cooling as the molten mass was moved rapidly to the surface to be erupted as lava. In this way, in a 1798 paper he melted and re-fused dolerite and was able to remark on the contrast between coarsely crystalline granite bodies, porphyritic whinstone (our dolerite/diabase) sills and finely crystalline basaltic lavas. He was aided in all this by the volcanic nature of parts of his homeland around the Scottish Borders and in the Midland Valley and by the granitic intrusions so common in the Scottish Highlands to the north.

Over the next decade, Hall turned his experimental talents to the action of both heat and pressure on the formation of limestone and marble from naturally occurring calcium carbonate starters (powdered calcareous shells, skeletons, etc.). Hutton had originally (and correctly) speculated that such rocks were the product of both elevated temperature and pressure. The latter disallowed simple calcining (as in limekilns) since the carbon dioxide (discovered by Hutton's close friend, Joseph Black, in the 1750s; his 'fixed air') produced would be forced to stay in place. Concerning the detailed direction of his experimental methods, here is the summary given by Peter Wyllie in 1999, a noted experimentalist himself and a modern biographer of Hall's career. Concerning Hall's 1812 paper he wrote:

> Hall completed more than 500 experiments using sealed gun barrels in various experimental designs with results that demonstrated the decomposition, recrystallization and melting of crushed calcite. Starting materials included chalk, limestone, spar [calcite crystal], marble and the 'shells of fish' [molluscs]. The calcite was enclosed in small vessels [Hutton's 'little crucibles']…placed at one end of a gun barrel…packed with refractory material…placed vertically or horizontally into a furnace. He [had thus] designed the first cold-seal pressure vessel, with the sample end of the barrel in the furnace and the cold end of the barrel outside…close by

a plug and folder to facilitate shutting and opening the barrel. Water was added to some experiments, to increase the pressure, and, as we know now, this facilitates fusion of calcite…The experiments confirmed that powdered calcite can be converted into hard, crystalline marble at high temperatures, as long as pressure is applied (representing a stack of overlying rocks)…He calculated pressures attained between 50 and 270 atmospheres, and temperatures attained between 20 to 51 on the scale of Wedgwood's pyrometers [a device originally invented by the noted potter, Josiah Wedgwood, to measure relative temperatures in his kilns].

Wyllie notes that it wasn't until nearly a century later that the high temperature of the beginning of the completely molten state (the liquidus) in powdered mineral samples could be accurately recorded – this by Norman Bowen in the sophisticated furnaces of the newly formed Geophysical Laboratory in Washington D.C.

Enter Norman Bowen

A few years or so after the time of the inauguration of the Geophysical Laboratory in 1907 (financed by the Scottish/American philanthropist, Andrew Carnegie), Norman Bowen graduated in geology and chemistry from Queen's University in Kingston, Ontario. He was the eldest son of a Canadian-born mother, his father a first-generation immigrant from London. Andrija Mohorovičić's discovery in Bowen's graduation year of 1909 subsequently raised great questions concerning the origin of the Moho and the role of the mantle in the genesis of basaltic magmas – topics that would fascinate Bowen in his long career. Much like the slightly younger Arthur Holmes, Bowen developed his own parallel course in research, not only from a deep and abiding interest in field geology and in mineralogy but also from developments in physical chemistry, notably of the application of Willard Gibbs's thermodynamics, which shed light on the fundamental controlling variables that determine the order of crystallization and composition of the mineral phases in cooling fluids.

During college vacations between 1907 and 1909, Bowen had worked for the Ontario Bureau of Mines mapping regional geology in remote northern Ontario, working around Lakes Larder, Abitibi and Gowganda – camping out and travelling alone by canoe most of the time. Bowen had caught the geology bug and was a pioneer explorer and recorder of a wonderfully diverse bedrock geology in what was natural wilderness, inhabited by Native Americans and abundant wildlife.

In the course of his field mapping he discovered several occurrences of thick horizontal bodies (sills) of dark-coloured, finely crystalline rock called variously dolerite (in England), whin or trap (in Scotland), and diabase (in North America). These layers of rock had originally been forcibly introduced in the molten state into horizontally stratified sedimentary rock, prising the layers apart. Their arrival changed the sedimentary 'envelope' in immediate contact with the magma due to an intense flux of heat, the cooling magma rapidly precipitating relatively small crystals at the 'chilled' contact where heat was being conducted away. What particularly intrigued Bowen were features that had a bearing on what was known as 'differentiation', the creation of younger, contrasting mineral aggregates in the dark-coloured diabase sills. This was indicated by the occurrence of small veins (dykes) and patches of light-coloured silica-rich rocks at their tops. His attempts to explain such features involved an interplay between the cooling and crystallizing magma and its interactions with water in the enveloping sedimentary country rock.

On graduation, Bowen entered the lists to achieve fundamental discoveries that lay at the heart of igneous petrology – he had written two original research papers, one awarded first prize by the Canadian Mining Institute and the President's gold medal, and the other appearing in the prestigious *Journal of Geology* associated with the University of Chicago. By the autumn of 1909, he had acquired the graduate's dream ticket – the prestigious study award of a British 1851 Exhibition Scholarship, whose bursary would support Bowen doing a doctorate at any place of his choice, undertaking research of his own choosing.

Initially, his wide reading led him to consider Norway, where a group of brilliant and innovative researchers (including J.H.L. Vogt and W.C. Brøgger) at the University of Oslo had published works on the igneous rocks of the aforementioned Oslo rift province. Unfortunately, Bowen's timing was all wrong, for Vogt was about to take up a prestigious new position at Trondheim and Brøgger had taken on an onerous administrative load. Correspondence with Norway relayed the difficulties of researching there (including the language) so Bowen gave up his efforts and, in the words of his biographer, H.S. Yoder Jnr, he:

> looked south for help from his fellow Canadian, the sparkling Reginald A. Daly at the Massachusetts Institute of Technology. It was from this inspiring teacher and while assisting him in the field, that Bowen received his title 'Daly bred'.

'Daly bred' at MIT

Daly imbued Bowen with a wider scope than just a field knowledge of igneous rocks. He introduced him to ideas concerning the probably simple causes of their great variety. One was his belief in the primacy of a glassy, basaltic mantle as the source of most magmas – two decades later this became a victim of Bowen's forensic criticism. On the chemical side it was C.H. Warren who taught that thermodynamics and physical chemistry were fundamental underlays to both the conduct of successful experiments and for their wider application to the origins of igneous rocks. In between summer fieldwork with Daly, Bowen gained a pre-doctoral position and began an experimental petrology research programme concocted by laboratory Director, A.L. Day, which involved melting and crystallization experiments on powdered mineral sample mixtures of exactly known compositions. These and subsequent experimental studies would revolutionize understanding of the evolution of igneous rocks.

The labs in which Bowen began his work were unique at the time. The high-temperature furnaces were controlled by banks of resister coils arranged to keep a constant and stable temperature, as measured by rare-earth thermocouples accurate up to 1755 °C. A rapid technique for quenching molten mixtures had been invented, and the compositions of the mixtures themselves accurately expressed in terms of a universal standard. A new petrographic microscope suitable for measuring finely crystalline and glassy proportions in the quenched samples was also available. The reasons for melting and crystallization of mineral phases from experimental mixtures were also understood.

Internal energies of minerals

Heat energy, the thermal equivalent of mechanical energy, is possessed by every substance and was the basis of thermal mechanics as first devised by James Joule. Later this was developed by Willard Gibbs to include the concept that heat energy produced by elevated temperature is caused by increased molecular movement and a trend towards disorder – the basis of what is known as kinetic theory (Greek; *konētikos* – to move). In fact, all substances have a certain definite amount of 'free' energy of formation for a given mass which is stored within the molecular structure of the substance in question – for example, in the mineral solids that concern us most from a geological perspective. This stored internal energy is termed enthalpy (Greek; *enthalpō* – warm) – for example, the hidden (latent) heat required from a mineral's surroundings to melt it completely. The greater the enthalpy the more difficult the pure substance is to melt, but the melting temperature is also lowered in solid mixtures. It also signifies the amount of energy required to alter a mineral by the process of weathering under earth surface conditions by aqueous chemical reactions – think of limestone dissolving or china clay (kaolin) forming from the solution of feldspar minerals and its re-precipitation as clay.

Experimental melting and crystallization

Readers may have had practical experience, when younger, of growing crystals from solutions of soluble chemical substances; for example, a saturated copper sulphate solution gradually crystallizes beautiful crystal forms. Magma itself is a molten silicate solution of elemental constituents – its crystallization during cooling to form the minerals that make up igneous rocks is a rather more complicated process. Experimental crystals were 'made' by Bowen in the laboratories of the Geophysical Laboratory during his pioneering work.

The procedure involved a systematic approach whereby small charges of certain proportions of crystalline constituents (gained from powdered pure natural specimens) were placed in sealed capsules made of non-reactive metal (such as platinum) and heated up by steadily increasing temperature increments – pressure in those early days was usually held constant at atmospheric, the isobaric condition.

As the required temperature was reached, the capsules were quickly removed from the furnace and quenched in a dish of mercury, the sample itself then removed for microscopic examination. Unaltered and unfused mineral powder meant that nothing had happened, neither change of state nor reaction – the sample ingredients had remained solid. With increased temperatures, identification of tiny amounts of glass (its type and composition identified by optical and chemical properties) in the quenched sample meant that melting had begun to take place, the melt having chilled to glass during the quenching. This defined the lower limit of melting for the mix of mineral constituents, a point known as the solidus. When the temperature was increased in successive intervals above the solidus, progressively more glass was recognized in the quenched samples until a point was reached when all the charge had melted and turned to glass. This marked the upper limit of the melting interval, known as the liquidus. Repeated experiments and analysis enabled plots to be made, as in the figure, of the boundaries of the various stages drawn as interpolated lines defining the solidus and liquidus for the melting system in question.

Application to rock-forming minerals

With his unrivalled experience spanning geology, physics and chemistry, Bowen quickly mastered theory, made intuitive decisions and established efficient laboratory practices. He turned his attention to the variation of chemical composition in certain minerals known to exhibit 'solid solution'. These have atoms in their molecular structures that range continuously in abundance between two end members, from zero to 100%. The 'solid' part of the apparently puzzling term – how can a solid mass at the same time be in solution? – refers to the fact that although a continuous variation in composition exists in the crystalline state of such minerals, this variation originated during crystallization from an evolving molten liquid. Solid solution and its implications for the evolution of igneous rocks was central to Bowen's life's work as an experimentalist. It also spawned major breakthroughs in the understanding of igneous intrusions and their field characteristics.

In the winter of 1912–13 he began investigations into the most abundant mineral in the continental crust – plagioclase feldspar – whose solid solution is between pure calcium- and pure sodium-bearing end members (Fig. 8.3). Melting temperature determinations indicated that anorthite changed state at around 1560 °C – it had a higher enthalpy than albite, which melted at around 1200 °C. Over three months he carried out melting experiments to determine the limits of the melting intervals at several different proportions of albite and anorthite, their melting temperatures decreasing from those of the end members. The results and their wider significance to the origin of igneous rocks were reported in an *American Journal of Science* paper

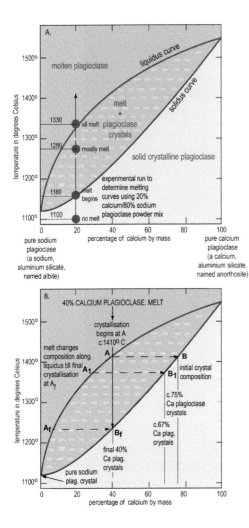

Figure 8.3 The melting (A) and crystallization (B) relations of the plagioclase feldspar solid solution series as determined by Norman Bowen in Washington in 1913. A. This plot of temperature versus chemical composition shows the complete liquidus and solidus curves with an illustration of their determination for an initial dry powdered sample charge of 20% calcium plagioclase and 80% sodium plagioclase. The results from four successive results for increasing temperature are shown by the solid circles. B. An illustration of the crystallizing paths for a 40% calcium plagioclase melt (after Gill, 2015). The initial crystals coming out of the melt at point A have a composition at point B of *c.*75% calcium plagioclase. This is more calcic than at point A – the melt is beginning to become depleted in calcium. With continued cooling and crystallization, the melt composition migrates down the liquidus curve towards A1 and Af.

of 1913, 'The melting phenomena of the plagioclase feldspars', the second of many experimental results published in that 'home journal' of Yale University during his experimental career over the next 45 years.

Bowen understood that examples of feldspars formed by gradual crystallization of a slowly cooling magma of a given initial mixture of calcium to sodium might be expected to abstract from a melt a larger proportion of the higher melting temperature end member, anorthite, and might continue to do so to a decreasing extent as the parent melt completely crystallized. Individual crystals would then vary in their composition from calcium-rich to sodium-rich, a phenomenon known as *zoning*, as seen in the plagioclase crystals sketched in Figure 8.1, the crystals having calcium-rich cores and gradually increasing sodium content outwards. The crystalline phase and its surrounding silicate melt would therefore change composition with time along

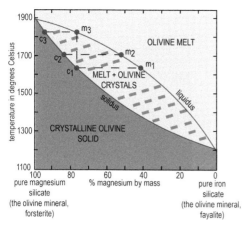

Figure 8.4 The melting and crystallization relations of the olivine solid solution series as determined by Bowen and his colleague J. Frank Schairer in 1935 (after Gill, 2015). As 75% magnesium solid olivine is heated up to point c1, olivine crystals of composition m1 begin to crystallize out – these are much less magnesium-rich than that of the starting olivine composition. Continued melting carries on along solidus to c3 with melts producing crystals of composition m3.

a pre-determined path – it would 'differentiate' (evolve). During natural melting or crystallization of mantle rock, differentiated magma might suddenly be separated from its parent, perhaps by crystals settling in the magma, as lava erupted at the surface, or might be intruded at shallower (cooler) levels in the crust, where it would constitute a separate rock body. In this way the separation of early crystals enabled later melts to have a different composition from the original melt (see Figure 8.3B) – a process known as fractionation or fractional crystallization. An analogy is provided by the fractional distillation and separation of crude oil into its constituent parts.

Another experimental result involving solid solution series was investigated experimentally by Bowen and J. Frank Schairer in a publication of 1935. This involved olivine, an attractive olive- to grass-green mineral, common in silica-poor rocks such as the basalts shown in Figure 8.1. It contains iron and magnesium as additional atoms to the silicon–oxygen tetrahedra that make up its crystal lattice. One end member is a purely magnesium-bearing variant, the other is purely iron-bearing. It was known to Bowen from published experiments that the magnesium form had a very high melting point of around 1900 °C, reflecting its high enthalpy. Pure iron-bearing olivine, by way of contrast, had a much lower temperature for change of state at 1200 °C. As he found in his plagioclase experiments, each lowers the melting point of the other – the trends in chemical composition during crystallization are revealed in Figure 8.4. Bowen used these, together with plagioclase feldspar fractionation, to explain the origins of distinctive basaltic rocks from the classic igneous province of the Scottish Inner Hebrides.

Two-mineral system with eutectic

In 1915 Bowen published experimental results involving two minerals that had no tendency to form composite forms or solid solution with each other and which would give no evidence for fractional crystallization. The minerals were

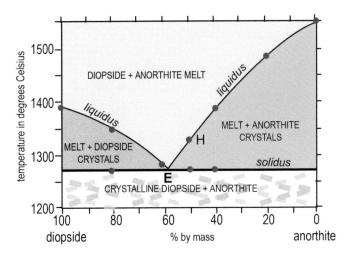

Figure 8.5 The melting and crystallization relations of the anorthite–diopside system to illustrate the occurrence of a eutectic point as determined by Bowen in 1915. See text for explanation.

the aforementioned calcium aluminium silicate, anorthite feldspar, and a calcium/magnesium silicate of the pyroxene family, diopside. The melting point of anorthite was at 1550 °C and of diopside at 1391 °C – each mineral in a mixture successively lowers its melting point. As Bowen daily plotted up his results, they gradually revealed the most elegant and simple form of a two-mineral (binary) phase diagram with a eutectic point (Fig. 8.5). This comprised two separate liquidus curves, one for each compositional mix, joining at a single compositional position of constant temperature – the eutectic. In this case the eutectic is at the constant temperature of 1270 °C. During crystallization of given mixtures, the two separate fields of anorthite crystals + melt and diopside crystals + melt join to define the eutectic field of simultaneous anorthite + diopside crystallization until exhaustion of melt and complete solidification.

In his 1928 classic book, *Evolution of Igneous Rocks*, Bowen pointed out the role of the eutectic in disallowing fractionation of chemically distinct magmas, unlike the solid solution examples discussed previously:

> If, during the early stages of crystallization, some of the anorthite crystals were removed or, say, settled to the bottom, no effect upon the course of crystallization would result therefrom. The composition of the last liquid to crystallize and the temperature of final crystallization would be exactly the same in the part to which the crystals had moved and in the part from which they had moved.

> Relative movement of crystals and liquid (crystal fractionation) could give rise upon complete crystallization to a mass locally enriched in either anorthite or diopside but to no other [compositional] contrast of one part

with another. Selective fusion of a crystalline mixture can, likewise, give no other contrast of one part with another.

Melting in the mantle

Evolution of Igneous Rocks defines what Bowen calls petrogenesis (*petros*: Greek for rock), the penultimate chapter of which moves up a scale from experimental and microscopical to the global. He writes in his usual wry style reserved for serious discussions of controversial issues:

> It is in many ways desirable to establish the connection of igneous activity with ascertained facts regarding the nature of the earth as a whole... Any system of petrogeny must, of course, be reconciled with geophysical facts, in so far as these are facts...A brief survey of the data and of some present-day conclusions in geophysical matters may be desirable, together with some suggestion as to their connection with the advocated system of petrogenesis.

Dismissing the relevance of the core, he reviewed the likelihood from seismic and heat flow evidence for a thick, solid shell (our mantle) made up of olivine-rich rock called peridotite (see below) that must underlie the crust. Furthermore:

> Having regard for the probable constitution of the whole earth, as indicated from the foregoing outline of data, we can see little chance of escape from the necessity of deriving all magmas *ultimately* from matter more basic than basalt and probably from peridotitic substance as represented in stony meteorites [their composition established by his colleague H.S. Washington in 1925].

Specifically, he politely but firmly demolishes the Daly scheme of a basaltic shell forming the lower crust:

> The preference is decidedly for crystalline diorite below its melting temperature in the layer where compressional wave velocities have the 6.4 km/sec velocity rather than for glassy basalt above its melting temperature in that layer...The elastic properties of the earth appear to reduce us to the necessity of considering the remelting of crystalline material. As already stated it is not likely that crystalline basaltic substance would usually give basaltic liquid in any process of remelting.

He posits that it was probable that basaltic magma was produced by the 'selective' fusion of a portion of the peridotite layer – the origins of 'parental magmas' for most igneous rocks – and that this 'suggest[s] a preference, therefore, for the production of these magmas by selective fusion of peridotite material caused by the release of pressure.'

Bowen was therefore well aware that magma generated in the mantle, such as the basaltic variety erupted underwater at mid-ocean ridges or as the vast outpourings during continental breakup known as 'plateau basalts', did not form from simple experimental mixtures. They must be derived from a mix of mineral constituents such as those occurring in the peridotite rock that makes up the mantle – olivine is the most common mineral but there are also pyroxene and garnet (Fig. 8.6). Such mixtures of mineral solids mean that mantle rock melts at lower temperatures than the simple mixes formed in the laboratory and described by the use of phase diagrams, the partial melt being the product of each of the mineral phases present.

The idea that a mantle peridotite 'mother rock' could generate the diversity of igneous rocks was perhaps Norman Bowen's greatest contribution. Later results from the Geophysical Laboratory and other institutions involving multicomponent melts under great pressure found relevance in the understanding of the process by which the rise of asthenosphere causes adiabatic (pressure-release) melting (see Chapter 15), which creates oceanic plate at mid-ocean ridges.

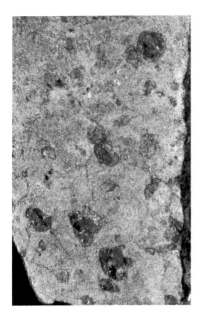

Figure 8.6 Photo of a 4 cm wide polished sample of fine Norwegian garnet peridotite, a rock established by Bowen as the major upper mantle 'mother rock' that gave birth to most others. The minerals present are: olivine (olive colour) in groundmass; larger crystals (phenocrysts) include garnet (pinot noir colour); orthopyroxene (grey); clinopyroxene (bright green). Photo reproduced from H.S. Yoder's classic discussion *Generation of Basaltic Magma* (1976).

PART 4

Stressful Stuff: Settings

Ernest Masson Anderson (1877–1960); Karl Anton Terzaghi (1883–1963); Ralph Alger Bagnold (1896–1990)

The planet, from core to troposphere, is entirely dynamic, in a variable state of stress. This is most obvious in our day-to-day world: weather forecast charts give us gradients of atmospheric pressure; wind gusts and river eddies transfer turbulent energy and sediment suspensions up and down; ocean currents transfer heat energy around their great 'conveyor belts'. More prosaic is the state of stress in the rocky outer layers of the planet, dependent partly upon the static gravitational attraction of its materials. Most obvious is the pressure of water columns in porous rock driving groundwater and artesian flow and the pressure of rock causing the buckling of mine tunnel supports and rock 'bursts'. A lot less prosaic, but also obvious, are the tremendous stresses caused by tectonics – devastating earthquakes and volcanic activity.

As we saw in Part 1, fault displacements generate earthquakes. Long after their 'death' – the time of the last known movement across them – they remain as witnesses to past displacement. It was the mechanical origin of the different kinds of faults that a young Scottish Geological Survey officer, Ernest Masson Anderson, investigated in the early twentieth century. Later in life, after early retirement from the Survey, he undertook explanations of geological features as an independent scientist. Most notably he specified the state of stress that allowed the intrusion of magma into the upper crust to form igneous features such as dykes and cone sheets. He was later considered the founder of modern structural geology.

Compared to the stresses governing the strength and failure of homogenous solid rock investigated by Anderson, the determination of stress in the materials comprising the outermost layers of the crust, as investigated by Karl Terzaghi, was a somewhat more complex task. These materials are the product of the interaction between Anderson's solid bedrock, colonizing life-forms, weathering and the depositional sedimentary agencies of water, ice and wind. Such varied influences make for a somewhat messy combination of unconsolidated sediment and partly- to wholly consolidated rocks. The stresses here involve both water *and* soil/sediment pressure, which combine to govern the behaviour of excavated loose sediment by defining the property of 'effective stress', the basis for the discipline of soil mechanics founded by Karl Terzaghi.

The turbulence of the blowing wind transporting desert sand and sculpting dune forms captivated the mind of Ralph Bagnold during desert exploration between the two world wars. Before his time, one of the mysteries of fluid dynamics was the role of viscosity – the resistance of a fluid to shear – in controlling the nature of fluid flow. The basic mechanics were first developed by Osborne Reynolds, introducing the concept of contrasting viscous and turbulent flow regimes. Natural surface flows transporting loose sediment and depositing it as stratified sedimentary deposits span the whole spectrum of what became known as loose-boundary hydraulics. By judicious use of theory, experiment and field studies, Ralph Bagnold laid down the proper dynamical rules that govern such processes – he was a founder of physical sedimentology and gave us the fundamental quantity now known as the Bagnold Number.

9

Ernest Masson Anderson (1877–1960) *Scottish mathematical geologist and independent researcher*

I have been unable to locate a portrait of Anderson, but see Figure 9.5 for him, aged 35, with a distinguished group of colleagues from the Scottish Geological Survey in 1912.

He related the types of rock failure during faulting to the mechanical stresses responsible by defining a system of stress distribution that varied according to the magnitude of three unequal principal stresses (1905). As well as original work on a host of other geological problems he later analysed the mechanical stresses responsible for igneous intrusions in relation to local and regional stress fields as disrupted by magmatic pressure from below (1936, 1942, 1951).

Faults in general

Faults are natural fractures that displace rock, sediment or soil masses along a plane of movement, as in the active examples cutting the land surface shown in Figure 9.1. They result from shear failure caused by forces transmitted through the earth's rocky and brittle outer shell of upper mantle and crust. The displacement (several decimetres to metres) generates kinetic energy which is dissipated by elastic deformation and the generation of seismic waves. As all readers will be aware, faults and faulting have taken off in popular 'jargon-culture' in the last decades. Recent media headlines include – 'Political Fault Lines Drawn'; 'Talks Run into Faultline'; 'Truss Faults Out' (no, that's invented!).

Richard Oldham knew that faults were the sole cause of earthquakes and, as we have seen previously, was able to locate and map out the culprit responsible for the 1897 Great Assam earthquake. Because there is usually no restoration of the pre-rupture state, faulted rock masses continue to bear witness to past tectonics by using the sorts of 'true cause' logic discussed in my Forewords. Old faults are best seen to advantage in stratified rocks where the strata have been displaced or, in the case

Figure 9.1 (opposite) A. Aerial view of a kilometre or so of the newly formed normal fault scarp during the 1959 earthquake along the east side of Pleasant Valley, Nevada, USA. The fault throws down the alluvium in the bottom of the view against the bedrock of the Tobin Range at the top. The fault had a total surface length of 15 km or so with a maximum ground displacement of around 1.5 metres. Source: Robert Wallace USGS. B. Oblique aerial view of the normal fault scarp and associated minor ruptures of the 1983 earthquake, Borah Peak, Idaho, USA. Source: USGS. Inset shows the author standing on the upthrown surface. The 1.5 m displacement affected a c.10 km^3 crustal block, giving a force at the base of the block of some 2.75×10^{14} N, equivalent to work done in a second or so of some 4.13×10^{14} J. C. The spectacular limestone scarp along the Pisia Fault, Gerania Mountains, Central Greece. The young Mack twins are leaning back on the lower, relatively fresh-looking c.60° sloping fault surface exposed by renewed faulting in an earthquake 30 years previously. D. The c.1.0 m horizontal offset of fencing produced during the great 1906 San Francisco earthquake preserved along a strand of the San Andreas Fault, California. The Jackson children stand on the fault itself, the sense of displacement being to the right; a dextral wrench fault.

Figure 9.2 (opposite lower) Simple block diagrams to explain the three possible modes of fault displacement – many faults have some combination of these 'end-member' motions.

of active faults, when cutting and displacing the land surface, these gradually reduce with time due to surface erosion.

Faults may be divided into three end-members (Fig. 9.2):

1. Normal faults displace material downwards in the direction of the maximum downslope inclination ('dip') of their faulted surface, commonly 50–60°. They are not 'normal', as in everyday language, but early geologists worked in areas where this kind of displacement was commonest. The magnitude of fault displacement from a particular earthquake may be several metres. Over time, with each subsequent earthquake, fault throw and the visible scarp of the slip plane grow vertically and to a lesser extent, laterally, in the landscape.

2. By way of contrast, faults that displace material upwards in the direction of the inclination of their fault surfaces are termed reverse or thrust faults – they show displacements opposite to those of normal faults. Such faults have surfaces that commonly, but not exclusively, show relatively shallow dips, < 40° or so, often very much less, or even horizontal for a class of faults known as overthrusts. Active reverse faulting was responsible for the Assam, Popusko and Murchison earthquakes featured in Part 1.

3. Faults that displace rocks horizontally *along* their usually vertical planes (their 'course' or 'strike') are termed wrench, transcurrent or

strike-slip faults. Such was the famous San Andreas fault responsible for the 1906 San Francisco earthquake. The direction of horizontal slip is obtained by either the left- or right-handed displacement of some surface marker (a fence, road, stream valley, etc.) as seen by an observer standing on one side of the fault looking over it to the opposite side (Fig. 9.1D).

As noted in Chapter 2 in connection with modern interpretations of the motion of the Pokupsko fault, Croatia, during the 1909 earthquake, seismologists can tell the nature of fault slip and therefore the kind of fault responsible for earthquakes remotely, by determining the pattern of the first arrivals of seismic body waves on seismogram records. Here it should be remembered that the three classes of faulting are but end-members of a continuum of behaviour, for many faults show a mix of relative movement across their surfaces – known as oblique slip.

Being Ernest: the early years

Ernest Masson Anderson's career with the Geological Survey of Scotland and his subsequent early-retired life as an independent scientist featured extraordinarily wide accomplishments: the interpretation of schistose lineations; crustal heat flow, melting and the origins and associations of magmas (with W.Q. Kennedy); advanced microscopy techniques. His most original studies, those that concern us here, were his early workings on the mechanical origin of faults (1905) and, 30 years later, the emplacement mechanics of dyke swarms, notably the cone sheets and ring dykes associated with former volcanic complexes in the Inner Hebrides and on the adjacent mainland.

He was born in Falkirk in the Midland belt to the Congregational cleric, John Anderson and his wife, Annie Masson, herself the granddaughter of a minister. He and his younger brother, Alan, were home educated till the age of nine, until in 1886 the family moved to Lerwick in Shetland, the father becoming minister there. According to a hospital interview in 1930 with his wife Alice, as a boy he was very upright and conscientious and did not associate with others of his age. This was partly to do with his mother's views of how little boys should look – long hair, Little Lord Fauntleroy suits, etc. Alice describes how he talked to her about quite an isolated childhood with no friends and little mixing at school in Lerwick (where he and Alan were bullied), being very focused on his studies and not liking or participating in sport. At the age of 13 he was sent to live in Dundee with his mother's father and went to Dundee High School – he was positive about this period in his life and did well at school, though by his own admission he was teased and tormented and at 15 made to wear braces for round shoulders, which depressed him. A year later he went to the University of St Andrews and did one year of an 'arts' course before beginning studies at the age of 17 for his B.Sc. in geology at the University of Edinburgh, where he graduated in 1897. Whilst studying hard he found himself entangled in internal debates with himself including unrequited love for a female cousin, the nature of materialism (reading Berkeley) and an exploration of Christianity, whose principles did not appeal to him, but nevertheless, experiencing what he described in autobiographical notes as a lifelong tendency to feel 'some compelling necessity to face what I did not like...It was not until I finished my university course that I got rid of this habit of mind.'

An indication of his natural bent towards quantitative science came with an MA with first-class honours in Mathematics and Natural Philosophy (Physics) the following year. After final graduation he entered the raw outside world via the teaching profession. However, he wrote:

> I was not successful as a teacher and resigned after six months, without any hint being given that I should do so. I made three other attempts within three years but always failed through inability to maintain discipline. I did not afterwards manage to extricate myself so satisfactorily as in the first instance.

It seems he was fired three times. Subsequently a family friend suggested he think about entry into the Geological Survey of Scotland. In typically thorough fashion he prepared himself for the entrance examinations for a whole year, passing in 1903 and beginning a 25-year career broken only by war service in 1916–1917, of which more later. In 1905, this highly sensitive and introverted young man, something of a social 'loner' without close friends and newly arrived at the Survey, published his first scientific paper. It was eventually acknowledged to be a classic in the annals of structural and tectonic geology. The paper was 'The dynamics of faulting' published in the Scottish capital's very own geological journal, the *Transactions of the Geological Society of Edinburgh*.

The 1905 paper

At Edinburgh during his MA studies, Anderson would have taken courses in the application of general physics to the mechanics of the deformation of solids, as originally set out by the early nineteenth-century French engineers Claude-Louis Navier and Charles-Augustin de Coulomb. It was Anderson's intention in 'The dynamics of faulting' to show how the different classes of faults are related to stresses in rock masses, part of the general problem of the deformation of solids by applied mechanical stresses by what is now known as the Navier-Coulomb approach.

As shown in Figure 9.3A, total stress can be quite simply defined by components at 90° to each other. Since all stress components are equal, then the situation so defined, the so-called stress field, is that of complete equilibrium – any tiny volume of rock is safely preserved and cannot be twisted, rotated, squeezed or expanded. If

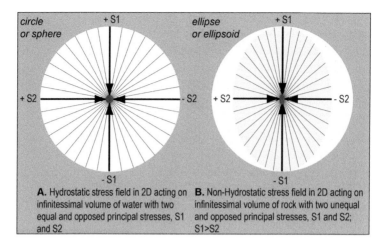

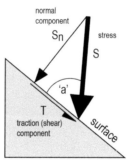

C. Normal and traction components of an applied stress over a solid rock surface; the ratio Sn/T equal to the tangent of angle 'a'

Figure 9.3 A, B. Diagrams to define 2D stress distributions acting upon tiny volumes of water and *in situ* crustal rock. C. Diagram to show the normal and tractional components of an applied stress to an inclined plane. As angle 'a' goes to zero, the stress is entirely tractional (shear); as angle 'a' increases to 90° the stress is entirely normal.

we are considering the situation in a still fluid, such static equilibrium is universal – stationary liquids cannot bear shearing stresses, a feature known since the days of Blaise Pascale as the hydrostatic condition. The tiny volume can be defined in 2D as a circle or in 3D as a sphere.

Rocks, unlike fluids, *can* bear shearing stresses but their state of stress may still approximate to the hydrostatic condition, known as a geostatic condition, when the principal stresses are equal. Anderson's problem was to devise a scheme whereby the commonly observed deformation of rock masses by faults could be physically accomplished by differential stresses. His solution concerned the magnitude of each of three principal stresses. He made the inspired assumption that the forces operating in real-world rock masses in the earth's crust – due to mechanisms such as unequal vertical loading or variable tectonic pushes and pulls – were not necessarily equal or of the same sign (as in contrasting tectonic compression versus stretching). So, he made each principal stress different, one stress as a maximum, another as an intermediate and the third as a minimum, still with each principal stress in one direction balanced by an equal and opposite stress. We recognize this as a non-hydrostatic stress distribution, defining a stress ellipse in 2D (Fig 9.3) and an ellipsoid in 3D (Fig. 9.3B).

The variable nature of such stress fields enabled Anderson to try to determine the position and orientation of the planes that will bear the *maximum* of any shearing stress. His reasoning was that such surfaces were the most likely potential sites for faults to break up the rock mass – planes that would determine the orientation of the faults that might arise from the competing magnitude and direction of the three principal stresses.

But how strong are rock masses and how can this strength be overcome by applied tectonic forces to produce rock fractures like faults? Anderson knew his Navier and Coulomb – they had proposed a criterion for the onset of brittle failure in solid materials that involved a critical stress needing to be applied for this to happen. This was based on the idea that failure under conditions of shear takes place along a surface when an applied stress is large enough to overcome the inherent strength of the rock mass. That this shear strength is highly variable can be widely appreciated by a familiarity with rocks as soft as crumbly sandstone or as tough and obdurate as finely crystalline basalt or dolerite.

There is also an additional factor involved in the Navier-Coulomb treatment – frictional resistance. For displacement to occur along a fracture at time of failure, then the two opposed surfaces of any initial fracture subject to stress must be dragged past each other. The magnitude of such a process is determined by the value of the product of stress normal to the slide plane times a measure of frictional resistance, a coefficient whose value might theoretically vary between zero for a perfectly frictionless contact and unity for total frictional resistance. A convenient way of thinking about this concept is to imagine the two opposed surfaces being pulled past each other as a sliding block along an ever more steeply tilted surface (Fig 9.3C).

All this can be expressed by the overall Navier-Coulomb criterion for failure *and* displacement of the form:

shear stress > internal strength + frictional resistance

It was now possible for Anderson to show by trigonometry that the planes of greatest shearing stress in a rock mass intersect parallel to the direction of the intermediate principal stress and inclined at certain angles to the other directions of principal stress (Fig 9.4). He also showed that the maximum sideways stress along a plane depends

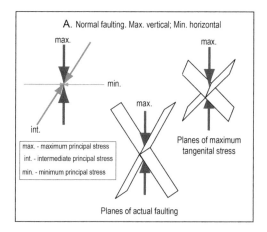

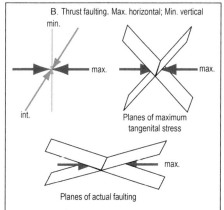

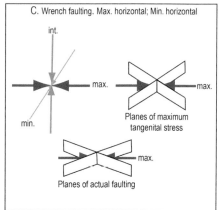

Figure 9.4 Versions of Anderson's 1905 diagrams explaining his mechanical theory for the origins of the three types of faults.

upon the magnitude of the difference between the maximum and minimum principal stresses and also on the angle between them, the latter giving a maximum at ±45°. Two intersecting fault planes would exist along these two planes of maximum tangential stress. In practice he recognized that the frictional nature of the rough rock surfaces being broken by faulting might act as a modifier to this ideal arrangement. The result would be that fault planes deviate from bisectors of the angle between the directions of maximum and minimum applied stress to form smaller angles with the direction of maximum stress. So, for faulting to occur, Anderson showed that there were three ways the three unequal principal stresses could cause a rock to fail and be displaced in certain directions. The sketches in Figure 9.4, re-drafted from his 1905 paper, illustrate the following descriptions.

First, when there is a relief of pressure in all horizontal directions (i.e., the rocks are pulled apart by extension or stretching) the vertical principal stress will be due purely to gravity and the intermediate and minimum principal stresses will be horizontal. In this case normal faulting will result if the strength and friction of the rock mass is exceeded sufficiently by the stress differential between the maximum and minimum stresses. We know today from studies of active faulting (Fig. 9.1A) that such faults and stress conditions exist during periods of lithospheric stretching forming rift valleys. Using a frictional coefficient of around 0.6, corresponding to an angle of internal friction of around 30°, the normal faults should show dips of around 60°, in line with many field observations.

Second, when there is an increase of pressure acting in all horizontal directions (i.e. the rocks are compressed) the minimum principal stress will be vertical and the unequal horizontal stresses will comprise the intermediate and maximum principal stresses. Reverse faults result when the strength and friction of the rock mass is exceeded sufficiently by the stress differential between the maximum and minimum stresses. We know today that such a regime arises naturally when plates collide, and that this type of compressional tectonics causes crustal shortening along reverse faults. Again, for a frictional coefficient of around 0.6, such faults should show dips of around 30°, as is commonly observed.

Third, when there is an increase of pressure in one horizontal direction combined with a decrease of pressure in the other, then a vertical shear plane exists with the principal and minimum stresses lying horizontal and the intermediate stress vertical. Under such conditions, if the strength and friction of the rock mass is exceeded sufficiently by the stress differential between the maximum and minimum stresses then a horizontal shearing (wrench) motion will result along the intermediate stress plane. Such faults are known today along the active axes of adjacent segments of mid-ocean ridges and along zones of oblique plate convergence at continental margins, as in the San Andreas fault.

The strength of rock

Anderson was naturally interested not just in the forces that drive rock deformation but also in the strength of those rocks to resist that deformation. However, he writes:

> It is very difficult to estimate what amount of tangential force will be necessary in order to produce actual rupture and so lead to faulting...Now, in the [experimental] case of a block crushed by pressure applied to two opposite faces, we are dealing with a single pressure, which we may denote by P...Assuming, then, that yielding does take place by shearing to begin with (and it is difficult to imagine what else could happen), the amount of tangential stress necessary to produce this shearing cannot be greater than P/2.

Referring to experimental work on rock fracture using small prismatic blocks, the force being applied to their ends, he quotes comments that:

> When a bar is pulled asunder [by tension], or a block is crushed by pressure applied to two opposite faces [by compression], it frequently happens that yielding takes place wholly or in part by shearing on surfaces inclined to the direction of pulling or thrust.

The correspondence of this result to Anderson's theoretical predictions was very clear. Experimental results by Ewings for common rock types like granite and basalt indicated that the critical applied stress for failure by fracture should not be greater than around 5 tons per square inch ($c.7.0. 10^6$ kg m^{-2}); for a hard sandstone around 2.5 tons per square inch ($c.3.5. 10^6$ kg m^{-2}).

Anderson then speculated as to how deep in the earth's crust such critical pressures might occur. For a hard sandstone resting under a column of overlying rock of average crustal density he calculates around 1.8 miles (2.9 km). This implied that below such shallow depths there should be continuous ('incessant') fracturing taking place – clearly an absurd result. It was clear to Anderson that to prevent this there must be two horizontal stresses acting in addition to the vertical stress considered from the uniaxial pressured experimental set-up of Ewings. At some depth these horizontal pressures might be expected to be about equal, perhaps approaching the value of the vertical load and giving rise to a geostatic condition. Here are his own words on this important deduction that would be answered by future studies of the deep state of stress in the earth using what we now call triaxial testing (all three stress vectors involved):

> At a depth of, say, 25 miles, the vertical pressure will be something very much greater than 10 tons per square inch, and so at this depth the differences between the pressures must be small quantities when compared to the pressures themselves. Thus there must be a condition of things, at great depths, similar in one respect to fluid [hydrostatic] pressure.

Faults that disobey Anderson's rules

It was Anderson himself who drew attention to what would become a major problem with the universal application of his stress theory for faulting. It concerned one of the most famous faults in Britain – the Moine Thrust of the NW Highlands of Scotland. Its discovery, mapping and (mis)interpretation had gripped the geological world in the 1870s and 1880s. Reputations had been made and destroyed concerning its true nature until careful fieldwork and deductions by Charles Lapworth led Ben Peach (who in old age became Anderson's mentor) and John Horne to reveal it as a low-angle thrust. They mapped the fault outcrop meandering around the mountainous landscape topography of Assynt, a testament to its essential horizontality. During its life as an active thrust it had carried Precambrian metamorphic rocks and an overlying Cambrian sedimentary sequence westwards over a 15 km tract of country comprising the very much older Lewisian metamorphic rocks.

Now, the nature of this fault alone might have scuppered Anderson's theory, for many low-angle thrust faults had also been discovered in the Scandinavian equivalents to the Scottish Highlands mountain belt. Yet, since these had a west-to-east transport direction, Anderson viewed them as complementary (mirror-image) to the east-to-west throw shown by the Moine Thrust. Yet he was unhappy with his analysis on account of the apparent mechanical impossibility of such low-angle structures. In his words:

> When we come to consider angles, however, the results of observation are less in accordance with our theory. Thus in Scotland the prevailing dip of the thrust-planes to the east is too low to have been produced by any purely horizontal pressure.

Anderson then rolled out an idea to solve this problem proposed to him by Peach – that both the Scottish and Scandinavian thrusts were indeed part of a complementary system but that they had both been produced on either side of a folded N–S trending mountain range whose vertical relief would have had the effect of changing the stress regime. As Anderson put it:

> the effect of the declivity of the ground on either side [of the putative mountain range] would be to tilt the directions of greatest pressure on either side...The planes of most likely thrusting would also be tilted, so as to become more nearly horizontal.

That this was an *ad hoc* (though ingenious) solution to a fundamental problem was clear, for there was no evidence for such a fortuitously situated mountain range, the tract of ground where it should be located conveniently buried under the waters of the North Sea. A convincing mechanical solution to the 'overthrust problem' involving subsurface fluid pressuring had to wait for over 50 years (see Chapter 10).

Reception of the 1905 paper

What did Anderson think about his own paper? He was sure of its significance, for in his previously unpublished autobiographical notes of 1930 he wrote:

> Apart from Survey work, the most important piece of work I ever did was contained in a small paper 'The Dynamics of Faulting'. This was published by The Edinburgh Geological Society. I wrote it not long after I joined the Survey, and it cost me comparatively little time. It was a wonder no one had done this piece of work before, as it was an extremely simple subject in spite of its importance. It is curious how much that is perfectly obvious has not yet been formulated by scientists. This was a case in point. Owing to the comparatively small circulation of the Trans Geol. Soc. Edin., this work did not become as well known as perhaps it should. An Edinburgh

Professor did it over again and read a paper to the Edinburgh Royal Society which he had to withdraw because I had covered the same ground.

We know what a later *eminence grise* of the Geological Survey, Sir Edward Bailey, thought about the man who left perhaps the most original and wide-reaching legacy of all the members of the Geological Survey during the twentieth century, with the possible exception of W.Q. Kennedy. In his magisterial history, *Geological Survey of Great Britain*, Bailey mentions Anderson but once, in connection with a 1915 investigation and analysis he had commissioned revealing the high aluminium content of an Ayrshire Carboniferous bauxite. He does spend many words of splendid criticism (and also some praise) of the man who, as head of the Scottish Survey in 1916, would become Anderson's nemesis, one John Flett.

Being Ernest 2: the Survey years

Anderson's life as a Survey geologist began well but was cruelly interrupted due to internal and external factors – within himself as he undertook arduous, solitary fieldwork and as the outside world drifted towards war. He wrote:

> My first few years on the survey were on the whole, happy years. For the first time I came into contact with men of real breadth of ideas and wide knowledge. The Survey work was very interesting, though I have never been able to conquer a feeling of tedium with regard to any sort of routine work. This was specially marked with regard to fieldwork ...it was always a relief to get back to the office.

This aversion to fieldwork was unfortunate, to say the least, in a youngish Survey geologist whose job demanded that he spent the summer half of each year doing usually solitary fieldwork in wild places. As we shall see, by 1909 his reaction to it was to have a debilitating effect on his life for a couple of years. Yet his winter situation at the workplace seems to have been hardly more convivial, for he writes:

> Until I became engaged to my wife [he married Alice Esson in 1915] I suffered very much from a feeling of friendlessness and tended to brood on it. This had not bothered me until the 1909 illness. The cleverer members of the Survey did not seem to desire my company much outside of office hours and I did not feel the same mental stimulus (in) the company of the less clever (who all the same were as clever as I).

On the plus side he had met Ben Peach, as he recollects in 1930:

> I had however a circle in which I was welcome in the household of the late Dr Peach, who was a retired Survey man, and had at one point been my

supervisor. I had a great regard for Peach himself. He seemed to understand natural phenomena in a wonderful way, though not exactly a critical scientist. He well deserved his FRS and his wide fame as a geologist. It was not however in discussions of geology that he showed to best advantage. He was a keen zoologist and botanist...... He could talk to shepherds and farmers about their callings in ways that impressed them....'He's a wonderful man Mr Peach' said one Grieve [Scots for farm overseer/bailiff]...'He kens mair aboot my own job that I do myself.' ...Peach's outlook on life had a great influence on me.

In 1909, during solitary fieldwork 'in an extremely lonely part of the highlands' (the mountainous terrain around Kinlochleven) he first developed serious symptoms of 'fear of open spaces, especially noticeable in an open boat and of heights' and which 'have been very troublesome with regards to my Survey work.' This agoraphobia (as it

Figure 9.5 Anderson with fellow Scottish Geological Survey participants posing outside the Inchnadamph Hotel on 1 September 1912 whilst on the British Association Field Excursion to Assynt. Those were the days – all-male, stiff white collars, ties, tweed suits, capes, the lot! As always, the smiling, athletic E.B. Bailey is pleasantly informal but the photo doesn't reveal whether he is wearing the trademark shorts. Back Row; W.L.P. McLintock, C.H. Dunham, G.B. Crampton, E.B. Bailey, J.E. Richey, M. Macgregor. Front row (left to right) E.M. Anderson, G.V. Wilson, J. Horne, B.N. Peach, G.W. Lee, W.B. Wright. Photo P008731 from the archives of the British Geological Survey, © UKRI, reproduced with permission.

is nowadays known) became acute enough for him to endure serious depression and to subsequently seek family help whilst undertaking field work, first in the company of his brother (whom he seems to have disliked – sporty, gregarious?) in 1912 and then of his 67-year-old mother during the field season of 1913 on the Isle of Mull.

To contrast with his almost allergic dislike of solitary fieldwork (it would return with a vengeance after the war), his attendance on the famous September 1912 British Association field excursion to Assynt must have seemed like a dream opportunity. For, just a few years after the 1907 publication of Peach and Horne's classic memoir on the geology of the NW Highlands, he found himself surrounded by the glitterati of a Europe-wide collection of geologists. They had come to worship at the altar of the Moine Thrust, whilst doubtless imbibing plenteously of an evening at their lodging place in the geofamous Inchnadamph Hotel, known thereafter to many visiting geologists. The excursion is noteworthy for us, since it was the occasion of a fine group photograph of him and his colleagues from the Edinburgh office of the Survey, including Peach, Horne and Bailey (Fig. 9.5).

After the troubled years of 1909–11, life seemed to have settled down for Anderson before the middle war years, for in April 1915 he was married to Alice Catherine Esson, daughter of a public-school teacher and thirteen years his junior, and with whom he had two surviving daughters (another died in infancy). They lived a sometimes uneasy life together until his death 45 years later. In March 1916, aged 39, newly married, he enlisted in the army as a private soldier (he wrote, 'from a sense of duty to his country'). This situation, a professional man of mature years with mental health issues enlisting as a private soldier seems surreal, yet such were the times. A further, personally challenging note comes from his recollection that in seeking permission to enlist, his 'chief' at the Survey advised him that he was not good officer material! This 'chief' was Deputy Director J.S. Flett whose deafness and rank disqualified him from military service – more of Flett later.

After infantry training, Anderson joined the 6th Battalion of the King's Own Scottish Borderers on the Somme, then shifted to the 2nd Battalion of the Highland Light Infantry where at Mailly-Maillet he experienced intense bombardment: 'perhaps the most terrifying experience I went through...nobody near was killed but 180 out of 450 of the group were killed or injured.'

Just after this, with another soldier, he was ordered by an officer to collect a wounded German soldier from a shell hole and take him to a medical dugout. They completed their admirable task but then got lost amongst the shell holes trying to find their way back to their own trench. The other soldier became annoyed at Anderson's slowness and left him wandering around until he was hit in the arm by shrapnel. Regaining his trench, the medical officer who examined him told him he had '...done it nicely – it was a sure blighty'.

Back in Scotland, the wound took six months to heal before he was subsequently transferred to the Royal Engineers as a sapper to train in Artillery Sound Ranging. However, a problem arose at once and he was demobbed soon after. Apparently, Flett at the Survey in Edinburgh had only agreed to Anderson joining up if he could be called back to the Survey at any time. Flett must have arranged this without telling Anderson, who had objected to the demob. This act added to his issues with his 'chief' after the war. Flett seems to have been a difficult man at times – E.B. Bailey's words of 1952 probably being an understatement:

> When Flett replaced Horne in 1911 a great change came over the Scottish Survey. In part this was due to the fact that the new Assistant Director was deaf, which tended to isolation of thought. On the publication question [the right of geologists to publish unofficial work in the open scholarly literature] he unfortunately went even further than Geikie, though from different motives. He claimed universal veto, whereas Geikie had admitted that unofficial work lay outside his jurisdiction.

The misunderstanding with his late-wartime recall seemed to hang over Anderson thereafter. In 1921 there was formal criticism of the time he was taking to complete a mapping project, perhaps due to a recurrence of his 'fieldwork problem'. Yet he was promoted to Senior Geologist in 1922 and had the important Schiehallion lineation paper published the following year. His first hospital admission in late August 1926 (after the fieldwork season), was to Craig House, a private psychiatric hospital in Edinburgh. By this time, he seems to have got more stressed about his relationships within the survey. He felt that Flett, along with Bailey, was overcritical of him and that others were being praised and promoted over him. He starts to seem rather paranoid and certainly his health was suffering – he was being prescribed anticonvulsant and strong sedatives by his family doctor before his admission. In 1928, he took (or maybe was invited to take) retirement from the Survey on grounds of 'ill-health' and had two further voluntary hospital admissions in 1929 and 1930. The last section of his 1930 personal account of his life is, in places, again a complaint about certain people at the Survey. He wrote no more autobiographical accounts of his immensely productive life in retirement and there are no records of further hospitalizations after 1930. Perhaps cessation of solitary fieldwork proved beneficial?

An independent scientist (1930–1960): on dyke intrusion

The Inner Hebrides islands and adjacent mainland that Anderson had helped to map in his Survey days host extensive basaltic igneous rocks of Tertiary age – sub-volcanic central complexes built up around formerly active magma chambers – seen to perfection on Ardnamurchan, Rhum, Mull, Staffa and Skye. These had been exhumed by erosion following crustal uplift over the past 60 million years or so, giving a unique opportunity to discover the physical and chemical processes that had gone on during their deep history as the foundations to active volcanic vents. They had spewed out awesome volumes of basaltic lavas at the surface, seen today as the so-called 'plateau

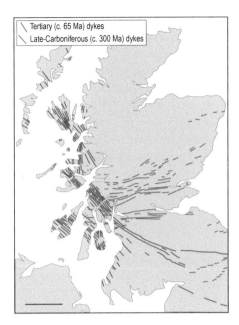

Tertiary (c. 65 Ma) dykes
Late-Carboniferous (c. 300 Ma) dykes

Figure 9.6 Scotland's diverse geology includes its status as one of the 'dyke centres' of the world. Here are James Richey's (pictured in Figure 9.5) classic 1935 and 1939 maps of its chief dyke swarms.

basalt' building blocks of thick lava flows seen so revealingly exposed on islands such as Mull and Skye.

Anderson was particularly interested in the immense numbers of dolerite dykes that he (on Mull) and his colleagues elsewhere had mapped out in now-classic investigations (Fig. 9.6). Such igneous-filled fissures had interesting but contrasting properties compared with fractures due to faulting. As he wrote in his 1942 book *The Dynamics of Faulting and Dyke Formation with Applications to Britain*:

> Dykes are, in the main, nearly vertical, and like faults they occur in parallel or subparallel systems, which have usually a wide lateral distribution. The two sides of each fracture are generally found to have moved apart in a direction normal to the fissure, so that the displacement is measured by the width of the intrusion...There is thus, in general, no lateral, and also no vertical dislocation, the latter fact being well known to coal-miners...

> It has been proved that faults should not theoretically be vertical, with the exception of one class, and this is also a matter of field experience. The class alluded to comprises the wrench faults, but it is obvious that dykes are not similar to these in character. There would in that case be double, and not single directions of alignment, and some evidence of horizontal movement would certainly be forthcoming. Dykes must therefore have been formed by some different mechanism, as they cannot, like faults, be explained by shear fracture.

Figure 9.7 A. Anderson's 1942 explanation for stresses involved in the intrusion of igneous dykes. B. An impressive dolerite dyke, one of a cluster of such dykes that successively intruded fragmented volcanic sediments and pillow lavas (left background) of remnant Tethys Ocean crust, Troodos Mountains of Cyprus (see Chapter 14). Exposures are along the gorge of the Akaki River near Klirou.

He goes on to deduce that the general characteristics of dyke swarms demanded that they should have formed in a state of regional tension but, as an addition, by the 'wedging action of the intrusive magma' at the time of formation causing 'a breakage which will extend the fracture' (Fig. 9.7). Also needed was a magma reservoir whose own fluid pressure is far greater than that of the transverse pressures in the country rock into which the dykes effectively squirt:

> The wedging effect of the magma will then be greatest along fissures which are perpendicular to the smallest rock pressure. These fissures will therefore tend to be selected. The magma will take the path of least resistance; that is, it will open up fissures along planes across which the pressure is least.

He goes on to ask two important questions concerning the implications of having dyke swarms radiating in many instances from particular magmatic centres and traversing scores, and in one notable case, to hundreds of kilometres away from them:

Figure 9.8 Coward's dyke, Namibia. Matchstick for scale.

It may be noted that there are certain kinematical, as well as dynamical, questions which must be answered by a theory of dyke intrusion. One may enquire, for instance, 1) what was the direction of advance of the edge of the fissure, and 2) what was the direction of motion of the magma?

In the first case the most distant example was the 400 km length of the Cleveland dyke of North Yorkshire as mapped out from its origins on Mull – it would seem to be the case of a horizontal advance, but it is impossible to be sure of the magma's direction of movement since some or all of it could have come up from below. It was, and is, extremely rare to find evidence for the sense of motion and/or the orientation of the shearing stresses that occur during dyke intrusion. My colleague at Leeds, the late Mike Coward, came across a fine example in Namibia and analysed it in his usual forensic way in an interesting paper published in 1980. The thin dyke (Fig. 9.8) showed that formerly spherical gas-filled cavities (vesicles) present in the magma source had been squeezed into ellipsoids by shear along the marginal boundary layers as it was intruded and cooled down to its solidus. By way of contrast, in the centre of the dyke where shear was minimal, the cavities remained near spherical. The flow profile derived from these strain markers by Mike was parabolic in form, suggestive of laminar (Newtonian) flow (see Chapter 11).

Central intrusions: the dynamical explanation

Cone-sheets and ring-dykes were named by E.B. Bailey and colleagues in the Mull geological memoir of the Survey in 1924. Intruded from a definite centre, they are common in several of the Tertiary volcanic centres systematically investigated by the Scottish Survey in the Inner Hebrides and adjacent Scottish mainland. They were also described from Ulster across the water and later on, equally spectacularly, at Spanish Peaks in southern Colorado, USA. These were figured by Reginald Daly (as his Figure 76) in *Our Mobile Earth* (1926), which Anderson and his brilliant contemporaries in the Scottish Survey, Edward Bailey, William Kennedy and James Richey must have read.

In all cases these centred intrusions form distinctive arcuate outcrops, arranged in series, round more or less identical centres of curvature. Cone-sheets are typically comparatively narrow (up to 15 metres) and incline in conically nested arrays at angles between 35° and 60°. Ring dykes are fewer and wider (50–2000 metres), and are either vertical, or dip *away* from their centres at steep angles. The two types are in several instances found to surround the same centre.

Anderson's elegant 1924 explanation of these fascinating igneous bodies (in the Mull memoir) was that they were due to stresses set up in fluid magma occupying a subterranean cauldron, supposing the domal magma to be under a greater pressure than that in the surrounding rocks. It will thus tend to lift its roof above the dome. Later, in the late 1920s and early 1930s, (with an extreme effort of application, as recorded in his autobiographical account) he worked out mathematical expressions showing how, with certain shapes of basin, and certain distributions of pressure along their walls, the stresses necessary for cone-sheet and ring-dyke formation arose. He published a highly original paper dealing with this in 1936, with a brief summary elsewhere in the following year. The principal directions of stress are given by one set of formulae (Fig. 9.9):

> Here the shaded margin may be taken to be a vertical cross section of the edge of the magma. If there is excess of pressure in the cauldron, there will be a relative tension which is greatest across the surfaces indicated by the fine firm [red] lines, with consequent production of cone-sheets in these directions. If there is defect of pressure in the basin, there will be relative tension across the surfaces whose intersection with the plane of the diagram is given by the broken [blue] lines. There has not, however, been intrusion of magma along these surfaces. The heavy [red] lines drawn at angles of about 25° to the broken lines represent the course of possible shear fractures. In their lower parts, at least, they correspond in inclination with ring-dykes.

It seems fitting to end this chapter with the comments of M. King Hubbert (see also Chapter 10) in his foreword to the 1972 USA facsimile reprint of Anderson's classic book published 30 years earlier:

> At the time when Anderson's book first appeared it had the distinction of being one of the few books, if not the only one, dealing with geologic structures and written by a geologist that was based on valid mechanical premises.

For aspects of his legacy through modern eyes, see the volume published after a conference at the University of Glasgow in 2010 held in his honour on the 50th anniversary of his death (Healy et al., 2012).

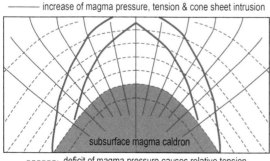

——— increase of magma pressure, tension & cone sheet intrusion

subsurface magma caldron

- - - - - - deficit of magma pressure causes relative tension
——— likely trend of resultant ring dykes

Figure 9.9 Version of Anderson's 1936 'dynamical explanation' for the origin of cone sheets and ring dykes.

10

Karl Anton Terzaghi (1883–1963)
Austro-Hungarian/USA soil engineer

Image of **Terzaghi** taken in 1926, aged 43. Unknown source. Wikimedia Commons.

He determined the effective stress existing in unconsolidated sediments (1925) and subsequently pioneered investigations of other fundamental properties and behaviour (consolidation) of engineering soil mechanic materials and the means by which they could be tested for stability limits in the laboratory and in the field.

Managing constructions on earth's surface

In the past, Chinese, Egyptian, Greek, Roman, Persian and Arabic engineers certainly knew the principles of hydrostatics that were later codified by Blaise Pascal. It was Karl Terzaghi who elucidated the basic mechanical principles of stress distribution in near-surface sediments and weathered bedrock and who established the regime of site testing and evaluation of safety that constituted the new science of soil mechanics – nowadays subsumed into the realm of engineering geology. He is quite universally regarded as the 'father of modern soil mechanics'.

Figure 10.1 Engineering soil mechanics often involves stability studies of excavations and built structures sited on highly weathered bedrock. Examples of the latter are shown here. A. Granite in Hong Kong excavation has been intensely weathered by the chemical alteration of its feldspar constituent under tropical monsoonal climatic conditions. The resulting brown, iron oxide-rich residue is a pasty granular mix with a low shear resistance, as shown by insertion of the shear vane. Photo courtesy of Steve Hencher. B. A similar situation on the Dartmoor granite, South Devon, England – note the less weathered remnants above the person's head. Here the china-clay (kaolinite) residue is a relic from early Tertiary times, when the Devon climate was much like that of Hong Kong today.

A geological perspective on soil mechanics, one that Terzaghi developed throughout his life since his student days, started with some simple general observations. The entire planetary surface is either drained by the tributaries of river catchments, scoured and scraped by glacier ice, or buffeted by desert trade winds. Vast areas of deposited sediment border rivers and continents with similarly large areas of wind-blown sands in the arid zone deserts. After atmospheric weathering, with the help of vegetation, various soil types develop on top of such unconsolidated veneers of sediment and altered bedrock – tropical weathering under monsoonal climates being a particularly efficacious destroyer of the fabric of even the most solid bedrock (Fig. 10.1).

Humans colonized these various and varied surfaces of the planet, and in due course they built their dwelling places, dug out water wells, walked out track lines, built roads, dams and reservoirs, excavated for ores, stone and fuels and constructed the communication networks that now cover much of the surface like a gigantic spider's web – footpaths, subways, tunnels, aqueducts, roads, canals, railways, viaducts, power lines and so on. The vast majority of the current 8+ billion humans living and labouring on this complex surface have their safety and prosperity dependent on the safe construction of the engineering structures that define their lives. For not only are there natural processes that pose existential threats to life and livelihood, like earthquakes, tsunami, avalanches and volcanism – these also arise from the unforeseen or negligent consequences of human engineering and construction.

Questions therefore need answering by the engineering geologist: will this proposed excavation for a foundation remain stable? Will construction of this new road or rail cutting collapse and endanger life? Is it safe to build on certain natural slopes? How steep can a containing wall or embankment safely be? Can a reservoir be safely constructed in a particular place? In all cases, subsurface water pressure and gravitational loading are the chief forces involved in controlling the mix of natural and human-induced hazards. These were the fundamentals that Terzaghi sought to understand during his long and pioneering career in soil mechanics.

The life

He was born in 1883 in Prague, then part of the extensive and long-established Austro-Hungarian empire, his father an officer in the Austrian army, the descendant of a long-established military family. The teenage son showed early signs of his independent nature by rejecting a military career in favour of a course in mechanical engineering at the Technical University of Graz, Austria. This gave him time to pursue wider interests in geology, philosophy and astronomy, not to mention his favourite outdoor sport of mountaineering. According to Arthur Casagrande, his chief biographer/obituarist, this mix included membership of a dissolute student society devoted to aggressive intemperance, but also with a bit of obligatory swordplay on the side witnessed by traces on his face of what he thought might be duelling scars.

Concerning the more intellectual side to his character, he loved essay composition and developed his geological knowledge at the expense of his chosen degree topic, to the extent that after graduation in 1904 and a year of army service, he found time to translate Archibald Geikie's *Outlines of Field Geology* and to accept a role as field geologist for the 1906 Greenland expedition – the famous *Denmark Expedition* of 1906–08. However, his chance of working with Alfred Wegener for two years on the ice cap was dashed by a serious mountaineering accident in the summer of 1906 – perhaps fortunately, not only for Terzaghi's safety on that sometimes ill-fated expedition, but also for the future development of soil mechanics.

On his recovery, the generous financial support of his grandfather ceased (his father had died in his boyhood), and he had to take paid employment as a civil engineer, working on numerous projects in Austro-Hungary, the Balkans and Russia over the next six years. He gained experience of person-management and an intimate appreciation of the complex interplay between natural landscapes, their geology and hydrology and the requirements for the safe and efficient construction of the newly introduced reinforced concrete engineering structures. With his geologically inclined mind, he was keen to make a proper scientific study of the consequences of earthwork and foundation construction such as the phenomenon of 'sand-piping' – the lateral eruption of sand from adjacent structures – and of settlement in response to loading.

To this end he spent what turned out to be a disappointing two years examining dam sites and their geological settings across the western USA. Returning rather depressed to wartime Austria, he had brief army and air force service before an unexpected and seemingly miraculous posting courtesy of the Foreign Ministry. This took him far away from war-torn Europe into the still-existing (just) Ottoman Empire, allied to the German/Austro-Hungarian cause. Here, in Istanbul, the engineering education of the Turkish state was being re-organized by his old professor at Graz, Philipp Forchheimer, who had recommended him for a teaching post at the Imperial School of Engineers. It was here and at the American Robert College after the war that Terzaghi first developed his lifelong studies of site investigations combined with experiments and the laboratory analysis of soil samples.

Granularity – soil and sediment matters

The sedimentary particles laid down as deposits by decelerating flows of debris, water and wind are all too familiar as a result of natural events like sediment avalanches, river floods and sandstorms. The residue from these comprises a variable mix of mud, silt, sand and gravel deposited during the complex de-mixing processes at work during sedimentation. What all sedimentary deposits have in common, though, is a more or less loose structure of mineral grains and particles plus the organic remains of animals and plants. The minerals are mostly of the silicate clan – rock-forming products from the weathering of rocks in some nearby or distant stream catchment.

At the surface, sediment has a range of spaces between its constituent particles. These voids or pores are the key to understanding the behaviour of unconsolidated surface and near-surface materials, for they are where water can accumulate to form natural subsurface reservoirs. For the purpose of analysis, natural sediment shapes can be simplified into uniform aggregates of spheres whose packing is defined by two contrasting end-members (Fig. 10.2). In 'cannonball' packing, each spherical layer fits snugly into its underlying neighbour, and the centres of adjacent spheres may be joined up in the mind's eye to define unit rhombs. The void space so created, the 'pore space', is the minimum likely one, calculated geometrically as around 26% by volume. By way of contrast, the maximum void space would occur if each layer could rest in tangent contact on its neighbour below and where the centres of adjacent spheres would define unit cubes. The theoretical space created by this cubic packing is around 48% by volume.

Natural sediment deposition clearly does not proceed with the careful stacking of spheres one by one on top of each other, but often in a rush of deceleration and dumping of sediment. Nevertheless, the two end members are essential guides to the range of magnitude of the void spaces produced in sedimentary layers. These spaces can be easily measured from core samples as a 'percentage rock volume porosity'. The value depends upon the processes of sediment transport applicable just prior to deposition, in particular the competing rates of winnowing versus fallout from transport. Two seaside and one kitchen example spring to mind. At the seaside we see beach sands stirred by repeated wave swash. These are firm to walk on because the

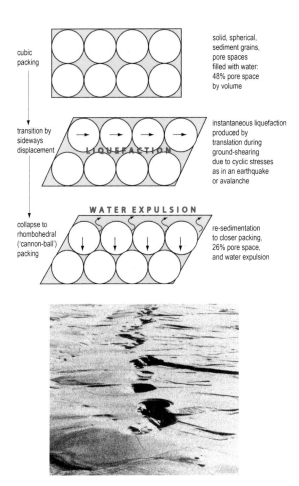

Figure 10.2 A. The chief modes of the packing of spherical sand grains and their re-arrangement to produce instantaneous liquefaction and expulsion of excess water and material picked up during extrusion. B. The result of sand/water expulsion as a fissure-line of *c.*2 metre diameter vented 'sand volcanoes' ('sand blows') after the 1989 Loma Prieta earthquake, California, USA. Image: United States Geological Survey.

constant to-and-fro action of wave motions causes close grain packing and therefore a relatively low porosity. By way of contrast, steepish coastal dunes receive grains from rapidly depositing avalanches that periodically move down them – they are soft to walk on because of their high porosity. The kitchen example is that almost any packet of dried beans or rice quickly emptied into a container can be reduced in volume by shaking or tapping the container. Such a kinetic process encourages grains to move into closer packing. Such considerations are important in what is known as powder technology.

Static forces

The most familiar example of static force, and one that serves well as an illustration of its general workings, is that of atmospheric pressure. Each morning's tap of the barometer in the hallway tells of small changes that we cannot sense directly. The weather forecaster shows us maps of such pressure and tells us which way and how fast the wind will blow, yet the static pressure itself in the house is not a vector like the velocity of the outside wind; it is a scalar quantity, having just a magnitude, and, most importantly in the context of this chapter, it acts equally in all directions.

The concept of hydrostatic pressure was Blaise Pascal's great discovery – that still water obeys all the rules of our atmospheric example. So, over the surface of the earth, every water-filled void joined up to its neighbours in the fabric of every sedimentary layer bears such a static pressure. Its magnitude at any depth is simply that borne by the weight of the overlying 'column' of water that rises to the surface – a tortuous but connected upward path. The density of water being so much greater than that of the atmosphere explains its far greater magnitude, yet it still acts equally in all directions at any one point. Variations in spatial pressure drive the rate and direction of subsurface flow in sedimentary layers (aquifers) in the same way as in the atmosphere.

It was Terzaghi's intuition in early-1920s Istanbul, that the state of stress (force per unit area) in the kinds of unconsolidated and often water-saturated materials he was concerned with as a practical engineer, for instance in the design of load-bearing structures, could only be calculated by a close consideration of the nature of subsurface stress. As it turned out, his intuition also applied to rock mechanics, enabling a solution to the major problem of the existence of low-angle to horizontal overthrusts in mountain belts that so concerned E.M. Anderson.

Effective stress

Although the water trapped in joined-up voids within porous, water-saturated sediment is in a hydrostatic state, it was evident to Terzaghi in a 1921 paper that the total stress acting on any unit area may also include a stress due to the mineral grains themselves. This is a contact stress transmitted at each solid surface that touches a neighbour in the framework of grains comprising the sediment mass. Its magnitude would be given at any depth below the surface by the submerged weight of all the grains above. Such a situation is described as a geostatic stress; its maximum value would be that appropriate for an entirely solid, void-free material.

Granular sediments like sands and gravels are undeformable at shallow depths, but in the case of mud-rich, water-saturated sediments the situation is entirely different, and this is where Terzaghi's proposal achieved its fullest development. Muds comprise tiny particles of clay minerals – these form the large and common group of so-called phyllosilicates ('leafy' silicates) that are the common weathering products of primary rock-forming minerals like feldspars and pyroxenes. Non-geological readers and potters will be familiar with at least one of these minerals, kaolinite or china-clay, the pure aluminium silicate. However, it is rare to find muddy sedimentary deposits that consist *solely* of clay minerals. Using the term 'mud' in a somewhat wider sense, we need to include the presence (revealed by electron microscope images) of tiny particles of pure silica (quartz) and other minerals, especially in glacially derived sediments.

It is easy to imagine that the gradual burial of muddy, water-saturated sediment will cause markedly different conditions in adjacent water-saturated sedimentary layers where fluid pore pressures are largely hydrostatic. To begin with, at very shallow depths, the saturated interstices between superimposed clay particles will remain in contact with each other and the surface, but gradually this hydrostatic condition will change as the weight of overlying muddy deposits increases. The clay particles defining the sedimentary framework will slowly be pushed and bent together by the weight force generated by the loading. The sediment pile gradually becomes deformed by this compaction and the overall porosity of the whole deposit lessens – the hydrostatic stress condition is replaced by an extra stress due to the loading of the overlying and compacting sediment pile. At the surface this process might become accelerated during deposition by sudden loading due to the dumping of thick deposits by flood events, or, in an engineering context, by the construction of superincumbent structures on top of porous sediments, such as buildings, bridge piers and dams.

Returning to the stable situation of a porous water-saturated sediment, we might imagine that any additional sediment deposited on top of such an accumulating deposit could only affect the states of stress in the already deposited material if it could materially alter the solid-to-solid grain contacts already established. Imagine an influx of coarse pebbles suddenly being dumped there – these materials would add weight to the whole mass of underlying sediment *and* to the pore water. The solid static stress would increase faster than the hydrostatic stress and act to try to compact the already existing grain mass.

The separation of grain-to-grain (geostatic) stress from hydrostatic stress is obviously necessary for engineering purposes, but also for explanations of the stress regime in the faulting of sedimentary strata. The grain-to-grain stress was termed 'effective stress' by Terzaghi, given by the simple formula:

Effective stress = Total stress – Hydrostatic (neutral) stress

He provides us with a neat experimental analogy by loading a sedimented loose sand layer within identical partially water-filled tanks in two contrasting ways (Fig. 10.3). In one tank he carefully deepened the overlying water column. This increase of water pressure at the top of the sedimentary layer increases only the hydrostatic stress in the grain mass below. In the second tank he loaded the sand layer with a solid, cemented layer of lead shot. The sediment below this solid layer now has no connection to the ambient clear water above, and so the weight force of the cemented lead shot sets up a non-hydrostatic effective stress.

Here are Terzaghi's own words describing his discovery some 15 years later. The details of his stress notation are omitted for clarity:

> The stresses in any point of a section through a mass of soil can be computed from the total principal stresses which act in this point. If the

A. Experimental conditions, A1 - A3

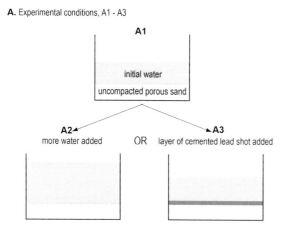

Figure 10.3 Terzaghi's simple but effective experiment to demonstrate the stresses involved in the accumulation and stability of layers of sediment grains. The layer of cemented lead shot represents a layer of compacted sediment, like a stiff clay or muddy sand mix that has much-reduced porosity and permeability.

B. Notional experimental stress and pressure measurements for A1 - A3

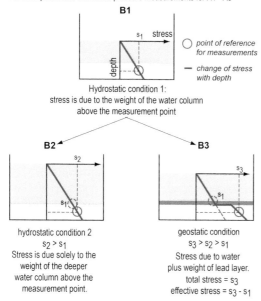

voids of the soil are filled with water under a stress, the total principal stresses consist of two parts. One part acts in the water and in the solid in every direction with equal intensity. It is called the neutral stress (or the porewater pressure). The balance represents an excess over the neutral stress and it has its seat exclusively in the solid phase of the soil.

This fraction of the total principal stress will be called the effective principal stress... A change in the neutral stress produces practically no volume change and has practically no influence on the stress conditions for failure…Porous materials (such as sand, clay and concrete) react to a change of [neutral stress] as if they were incompressible and as if their internal friction were equal to zero. All the measurable effects of a change of stress, such as compression, distortion and a change of shearing resistance are exclusively due to changes in the effective stresses. Hence every investigation of the stability of a saturated body of soil requires the knowledge of both the total and neutral stresses.

In fact, the Terzaghi concept was not a new one, but it had never before been isolated, named and analysed from the point of view of the principles of mechanics, just as faulting had long been recognized, but not explained until E.M. Anderson's work. As was pointed out by the late Alec Skempton, a distinguished engineering geologist at Imperial College, London, a more flowery, but essentially correct, description of the state of total stress and its subdivision into hydrostatic stress and geostatic stress was presented by Charles Lyell in his *Elements of Geology* as long ago as in the edition of 1871:

When sand and mud sink to the bottom of a deep sea, the particles are not pressed down by the enormous weight of the incumbent ocean; for the water which becomes mingled with the sand and mud resists pressure with a force equal to that of the column of fluid above. The same happens in regard to organic remains which are filled with water under great pressure as they sink, otherwise they would be immediately crushed to pieces and flattened. Nevertheless, if the materials of a stratum remain in a yielding state, and do not set or solidify, they will gradually be squeezed down by the weight of other material successively heaped upon them, just as soft clay or loose sand on which a house is built may give way. By such downward pressure particles of clay, sand and marl may be packed into a smaller space, and be made to cohere together permanently.

Compaction and consolidation

Modern soil mechanics engineers call the overall combination of sediment compaction and water expulsion 'consolidation'. Its particular state in sedimentary deposits is a fundamental issue in determining safety limits during excavations, for if it happens that local factors have prevented water expulsion, then a sudden and large pressure change arising from excavation or construction might be capable of causing structural failure. The generation of effective stress and the concomitant deformation and water expulsion by 'sand-piping' around the margins of earthwork and concrete dam margins built upon clays and interlayered sands, and their resultant failure, was the chief challenge that faced Terzaghi at the onset of his experimental investigations in 1920s Istanbul. Modern geotechnical procedures during site investigations involve a

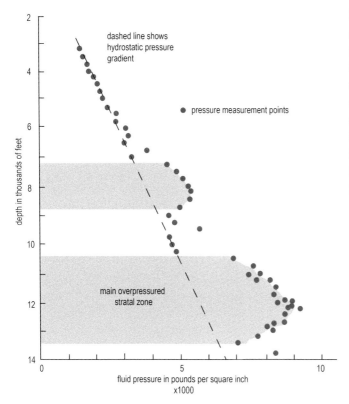

Figure 10.4 To show the rapid increases of subsurface pressures in sandstone sediments overlain by compacted mudrock sediments measured in a hydrocarbon well from Louisiana, USA (redrawn after Schmidt, 1973).

variety of drilling rigs and pressure-measuring kit used for the analysis of the yield-bearing strength and other characteristics of sedimentary materials that make up the subsurface framework upon or within which construction takes place.

The overpressuring solution to Anderson's dilemma

Following E.M. Anderson's lead and pioneered initially in the European Alpine chain, by the 1950s low-angle overthrusts and associated folds (nappes) were known from mountain belts all over the world. Exploration for deep oil and gas deposits in the Rockies of Wyoming and elsewhere had also led to a detailed knowledge of thrust fault geometry and their associated folds. We now know, thanks to the work of M. King Hubbert and William Rubey in the 1950s, that such faults arise during rock deformation by the interaction between rock stress and the stress due to water trapped in the pores and cracks of rocks by 'overpressuring' (Fig. 10.4). This leads to a reduction of rock strength that enables fracturing to occur at lower angles to the applied stresses than predicted by geostatic pressure alone. Such developments arose from the concepts of 'effective stress' as introduced by Terzaghi.

Earthquake liquefaction and Reynolds' dilatancy

A particular kind of dangerous and instantaneous failure in loosely-packed, water-saturated sands commonly occurs during the seismic activity associated with strong earthquakes – an association first noted and photographed by Richard Oldham in Assam (Chapter 1). A dramatic instance of such failure was given to me during a field excursion to a seismically active part of central Greece along the southern shores of the Gulf of Corinth. Our driver, Dimitrios, arriving on the outskirts of his native village of Skinos, pulled the vehicle over and beckoned us to get out and take in a particular view. 'I want to show you something', he said. The locality was part of a narrow, sandy coastal plain with the mountains of Helicon in the distant north, the Gerania mountains just behind us and the blue waters of the Gulf of Corinth shoreline nearby. Waving his arms, he said this was where his family had lived until the end of February 1981 when he was a young boy. So where was your house, we asked, puzzled at the absence of a dwelling place. He pointed downwards: 'It sank up to the first floor, disappearing as we all ran for our lives!' he said. Dimitrios and his family had survived a magnitude 6.5 earthquake along the normal fault system that marked the boundary of the Gerania mountains with the coastal plain we were standing on. During the earthquake 22 people died and hundreds were injured in adjacent mountain villages, some more than 20 kilometres away. But why should a house on the coast sink like that?

The reasons become clear when one studies the efforts of one very mathematically inclined experimentalist, Osborne Reynolds. He was an Irish-born prodigy from an old Suffolk family, appointed, aged 25 after a Cambridge education, to a professorship of engineering at the University of Manchester, the first such chair in England. In September 1885, Reynolds reported his experimental and theoretical results concerning the behaviour of water-saturated sands in a lecture to the British Association at Aberdeen. He subsequently wrote up his main conclusions, published in the *Philosophical Magazine* later in the year as 'On the dilatancy of media composed of rigid particles in contact, with experimental illustrations'. There was a new word in this title: dilatancy. It was coined by Reynolds to explain changes in the packing arrangement of sand subject to shearing motions, most notably in the change of arrangement of packing modes noted previously. Reynolds had noticed the effect of shear in the everyday experience of a person's walking upon water-saturated beach sands. In his own words:

> A well-marked phenomenon receives its explanation at once from the existence of dilatancy in sand. When the falling tide leaves the sand firm, as the foot falls on it the sands whiten, or appear momentarily to dry round the foot. When this happens the sand is full of water, the surface of which is kept up to that of the sand by capillary attraction; the pressure of the foot causing dilatation of the sand, more water is required, which has to be

obtained either by depressing the level of the surface against the capillary attraction or by drawing water through the interstices of the surrounding sand. This latter requires time to accomplish, so that for the moment the capillary forces are overcome; the surface of the water is lowered below that of the sand, leaving the latter white or dryer until a sufficient supply has been obtained from below, when the surface rises and wets the sand again. On raising the foot it is generally seen that the sand under the foot and around becomes momentarily wet; this is because, on the distorting [i.e. shearing] forces being removed, the sand again contracts, and the excess of water finds momentary relief at the surface.

Applying these principles to the effects seen during strong earthquakes, intense to-and-fro shaking motions due to the passage of seismic waves cause the momentary disruption and separation of the grain fabric of water-saturated sand deposits. As in Figure 10.2A the sands are liquefied – they become 'quick' – and as the grains fall back into a tighter packing, water issues out under pressure. Should this pressure become critical, then the water emerges at the surface in jets of sand-laden fluid. This is the origin of the extensive and ruinous sheets of erupted sand and sand 'volcanoes' (Figure 10.2B) made visible after great earthquakes affecting alluvial lowlands like that of the Brahmaputra valley seen by Oldham during 1897. And Dimitrios' house? Well, any building or edifice whose foundations rest on such liquefied sand masses will simply sink, its interior rooms filling with sediment and water as it does so.

11

Ralph Alger Bagnold (1896–1990) *English professional soldier, desert explorer and independent scientist (sedimentology and geomorphology)*

Bagnold in 1929, aged 33, on an early desert expedition, as featured in his 1990 autobiography. He is sitting by his stripped-down Model-T Ford. Image courtesy of University of Arizona Press.

He determined the magnitude of the stresses acting in blowing wind, flowing water and water waves, developing the fundamentals of the mechanics and rates of sediment transport and bedform development as related to flow power. He coined the term 'loose-boundary hydraulics' and introduced an entirely original explanation involving Reynolds' accelerations for sediment suspension by air and water currents (1930s–1970s).

In Reynolds' footsteps – the fluid physics of loose boundaries

For many years, viscosity, the fluid property of resistance to flow (think runny honey vs water), was viewed by physicists as an inconvenient obstacle to their efforts to express fluid movement in a mathematically precise way. So, they invented 'ideal' fluids that had none of the real-world properties attributable to viscosity. All this changed in 1883 when Osborne Reynolds published the results of carefully designed water-flow experiments in a purpose-built laboratory channel. He identified one type of flow that was quite uniform in space, with flow markers sliding lazily along parallel paths with a general lack of mixing. This he termed 'direct' flow, now known as laminar flow. At some critical value of increased velocity, flow behaviour seen at any point changed utterly, becoming unsteady, with twisting (vortex) motions causing intense fluid mixing. This kind of flow was termed 'sinuous' by Reynolds; it is now known as 'turbulent', dominated by rising and falling three-dimensional eddies that thoroughly mix up the fluid.

To explain his observations Reynolds came up with the idea that a certain ratio of fluid properties would govern such radically different flow behaviours – it involved a simple division of the multiple of flow depth, flow velocity and fluid density (the inertial force) divided by fluid viscosity (the viscous force). Since his death in 1912, this ratio of inertial to viscous forces in fluids has become known as the Reynolds Number (Re) in his honour – one of the first non-unitary (dimensionless) numbers used in physics and fluid engineering. Sculpturing of Earth's weathered surface by flowing wind, water and debris (also by volcanic ash flows and lava flows) may involve both types of Reynolds' flow patterns. They are an ever-present contribution to global recycling, the flows transporting detritus into depositional sinks like the oceans and across the trade-wind deserts of the world. By way of contrast, water no longer flows on the Martian surface, but its wind still blows strongly. The resulting drifts and dunes partly cover the immensely old deposits of an Archaean waterworld that nowadays entices new voyages of exploration after the excitement of the first Viking missions of the late 1970s.

Reynolds' further work published in 1895 on the nature of turbulent accelerations occurred just eight years before the first successful aeroplane flight was undertaken by the adventurous Wright brothers. Soon the nature of wind flow over aerofoils, and in water flows past ship hulls, became an essential field of study. Such ideas define the modern discipline of fluid mechanics and the concept of boundary layer flow. They were brought into the mainstream of general physics by the subsequent work of Ludwig Prandtl in Germany, Theodor von Karman in Hungary/Germany and Geoffrey Taylor in England.

Enter Ralph Bagnold

The application of boundary layer studies to the understanding of earth's surface sediment transport was pioneered in the 1930s by Bagnold, who, after army service in World War I, became an engineering student under Taylor at Cambridge. He then re-enlisted, later retiring in the 1930s with the rank of Major. Whilst still on army service with comrades in the late 1920s and early 1930s, he spent his annual leaves on pioneering vehicular explorations of the Western Desert of North Africa. He wrote vividly about these inter-war experiences and adventures in *Libyan Sands*

(1935), a delightful book recently re-issued as a modern classic. He later undertook field studies of wind-blown sediment transport during desert sandstorms, reinforced by laboratory wind tunnel experiments. This pioneering work was written up as one of the great classics of experimental geology, *The Physics of Blown Sand and Desert Dunes* (1941). After active service in World War II he worked closely with the United States Geological Survey on the transport of sediment by water flows.

Bagnold came from a military family on his paternal side, a tradition stretching back to the early nineteenth century. His father was a high-ranking member of that elite corps of the British Army, the Royal Engineers, whose origins date from the long wars against Bonaparte. Then, as now, such soldiers had to possess vital skills in signalling and engineering as applied to field situations that might have more than a frisson of accompanying danger. Bagnold's father possessed such skills, notably in telegraph, telephone and radio communications, and practised instrument-making in the Bagnold home in a well-set up workshop featuring a precision lathe and other useful kit. Later in life Bagnold would manufacture all his experimental fluid dynamical apparatus himself, from a plywood wind tunnel to coaxial rotating cylinders and suspension tubes.

Bagnold and his older sister, Enid, had much-travelled childhoods as their father was posted to various commands, living in Jamaica for several years. Both children

Figure 11.1 Bagnold in uniform as a young lieutenant of Engineers at home on leave from the battlefields of Third Ypres (Passchendaele), Belgium in 1917, with his mother a blurred presence in the background. Photo by sister Enid.

My brother Ralph on leave from Ypres. My mother in the background were encouraged to be independent and resourceful, qualities first seen in the rebellious and frank attitudes of Enid to strictures put on her social activities by their sometimes irascible father. Seeking an independent life as a writer she entered the world of literary London, eventually becoming a celebrated and highly individual playwright (*National Velvet*, etc.). Bagnold himself was educated at Malvern College until, on the brink of choosing a university course, the outbreak of war intervened – in 1915, aged nineteen, he enlisted. He spent three years as a field engineer in support of frontline troops in the trenches of the Western Front at the major battles of Ypres (Fig. 11.1), the Somme and Passchendaele.

From Cambridge to *Libyan Sands*

After the horrors of the trenches, post-war opportunities arose for the education of the 'missing generation' called up in the war years with the enactment of ex-officers' right to access university courses. He chose to study engineering and related subjects and, like Patrick Blackett, was accepted by the University of Cambridge. He studied the Engineering Tripos, graduating with an MA after only two years. In his autobiography he writes delightfully of the experiences and freedoms of post-war Cambridge and especially of the opportunity for peaceful foreign travel, which he was now able to enjoy to the full. He chose to resume his army career in 1921 and was posted to the 5th Division Signal Company, serving in Ireland during its Civil War, where officers were required to carry loaded sidearms if they walked out in civvy clothes during leave. On the transfer of army signal units from the Royal Engineers to the new Royal Corps of Signals, he was sent to its Catterick Training Centre as the Instructor in electricity, later as Chief Instructor.

In 1926 he was posted to Egypt, serving there for two years. The social attractions of peacetime Cairo meant little to Bagnold and his circle of youngish, single junior officers. United by a love of the outdoors and especially the almost boundless opportunities opened by the availability of personalized motorized transport, they mounted increasingly ambitious expeditions into the almost unknown interior of the Western Desert. Together, the small group of four to five friends served their exploration apprenticeships in planning routes and terrain evaluation using military maps from Allenby and Lawrence's celebrated campaigns with the Camel Corps against the Turks in the First World War. They carefully planned the location of fuel and provision dumps, particularly of water, carrying essential vehicular spares and so on. Initially they travelled along well-known routes eastwards from Suez into Sinai and along the still untravelled rugged mountains bordering its coastline and the Jordan rift depression. This was not just 'off-road' stuff, as there were no roads in the first place, only lines of markers – ancient camel skeletons along the caravan ways that joined up to make the ancient trading routes into and out of the Nile civilization to the west. The trail from Giza to the Sinai copper mines across the Gulf of Suez, for example, is now known from papyrus fragments to have existed for at least 5000 years.

From 1928 into the mid-1930s he was posted to diverse locations in the British Empire including the dangerous and challenging North-West Frontier of India in 1928–31 where he commanded the Waziristan Signals. It was during annual leaves from India that he returned to his beloved desert landscapes in organized expeditions with his old comrades in 1929, 1930 and 1932. Gradually their horizons extended westwards as they explored routes between legendary oasis settlements once visited by Alexander the Great. Passage over the stony surfaces of the rocky desert regions changed into daring attempts to force their stripped-down machines with super-deflated tyres up, over and down the steep east–west slopes (loose sand can stand at angles of 30° or so) of the chains of linear (seif) dunes that mark the un-navigated beginnings of the Great Sand Sea. Compass navigation from the metallic chassis of their cars proved difficult, and to avoid delays in having to stop and sight away from the vehicles he designed a sun compass that could be quickly and accurately read whilst on the move.

Libyan Sands was a tribute to the men and machines that made the explorations he described in successive reports published in 1931 and 1933 in *The Geographical Journal of the Royal Geographic Society*. The book itself was written for general readers who might have an interest in the landscapes, history and cultures of the Western Desert. His writing style was both informal and informative and the whole project with its

Figure 11.2 A desert landscape like those familiar to Bagnold and his colleagues during their explorations of the Libyan interior in the 1930s. Here is a procession of curved-crested (barchan) dunes, 5–10 metres high, that have been shed from a sand mountain (draa) to the right, making their way across a coarser-grained and immobile alluvial plain. Southern edge of the Grand Erg Occidental, Algeria. Tyre tracks in the left foreground for scale. Bagnold explained the nature and origin of such dunes in his *Physics of Blown Sand and Desert Dunes*, published in 1941 as his troops of the Long-Range Desert Group were patrolling such terrains in converted Chevrolet trucks.

perfectly drafted maps is infused with a sense of mystery, respect and admiration for the various cultures that had once, even in pre-history, inhabited and travelled there. His descriptions of the natural desert landscape were imbued with a developing knowledge on his part of the mechanical reason for desert dynamics – the wind systems funnelled by stark mountain ranges, the different kinds of dunes (Fig. 11.2), the driving particles in waist-high and lower sandstorms, the 'singing' of the desert sands in the lee of great dunes as their sand avalanches rolled down them. Then there was the informality and release enjoyed by the youngish soldier-explorers as they left the confines of their military lives in peacetime Cairo – camping exhausted under the stars after arduous days digging their vehicles out from sand drifts or passing long sections of galvanized steel under wheels in snail-pace progress through the most difficult ground.

An independent scientist

After a period in England as Chief Instructor at the School of Signals at Catterick, in 1933 Bagnold was posted to the Far East as Officer Commanding Signals, China Command. During all this time in the army he had busied himself in his spare time by researching and writing up not only his geographic explorations but also his thoughts and observations on sediment transport in the form of original research papers in the *Proceedings of the Royal Society*. These were transmitted in the first instance via Taylor, but after his election to Fellowship in 1944 he was able to access this publication outlet in his own right.

G.I. Taylor had by that time become the most eminent and original fluid-dynamicist of his generation with his discoveries born of a lethal combination of mathematical elegance, informed theory and practical experiments of great instrumental originality. His varied background and research interests meant that he could read and appreciate Bagnold's novel contributions gained from atmospheric boundary layer theory. This had been outlined in the first decades of the twentieth century by Taylor himself, by Ludwig Prandtl at Göttingen and his student Theodore von Kármàn (who later escaped the Nazism espoused by Prandtl to become a US citizen).

On 11 February 1935 Bagnold read a paper to an afternoon meeting of the Royal Geographical Society in which he summarized some of his thoughts on wind-blown sediment transport. The paper was written and published just before his retirement from the army and before he began a long-lasting and fruitful series of laboratory experiments at Imperial College, London. Bagnold begins his paper with what would be a *leitmotif* throughout his long career as an independent scientist – the need to establish general principles before attempting to explain natural phenomena – a trait he possessed in common with the majority of the other pioneers dealt with in this book. He wrote:

> It seems to me that there are two sets of problems to be solved, and that before any of the fascinating questions connected with the shape and

regularity of the great desert sand-dune systems can be tackled seriously we must solve the more fundamental problem of why and how sand accumulates at all, instead of scattering haphazard over the country. In other words, we must try and find out the mechanism of the interaction of the wind with the individual flying sand grains. Only after that can the second set of problems be tackled, namely the interaction of the shape, size, and repetition distances of the dunes themselves with the sand-driving and sand-depositing wind, and therefore with the equilibrium conditions which must give rise to that shape and size.

Dealing with the more general problem, Bagnold insisted on a distinction between the wind-formed dust clouds due to suspension and atmospheric turbulence and the movement of sand grains driven closer to the desert floor in what we now know as 'bedload' transport:

The whole appearance of the air during a sand-driving wind is very different from the pictures one sees of vast rolling masses of dense cloud in Indus Valley, the Sudan near Khartoum and elsewhere, which are erroneously spoken of as sandstorms. In the Libyan Desert, except during a severe storm, the visible cloud rarely rises higher than one's shoulders, and the sky above is almost clear. The cloud may be very dense, the crests of dunes may alter shape as one watches, the whole surface may be removed from round one's feet as one stands, but the top of the cloud remains within a metre of the ground as a clearly defined surface: a moving carpet.

Defining the ratio between the force of gravity and the pressure force acting on a grain, he was able to calculate the susceptibility of any grain of a given size to be moved by the vertically directed turbulence of an air flow, a ratio of value one or more being the criterion for the steady upward motion. Very high values of susceptibility for smaller grains implied that they would always be transported aloft in suspension load, whereas common sand grains of average diameter could only be kept suspended by the vertical eddies of turbulence when 'the necessary supporting upward air currents...[are]... equal or greater than the terminal velocity of the grain'.

The carpet-like nature of the sand grains in transport and the lack of evidence for sufficiently high vertical turbulent velocities for suspended transport meant that another mechanism must be sought for the maintenance of bedload transport. From a combination of physical intuition, simple laboratory experiments and field observations of the contrasting nature of relatively small fine-grained sand ripples and larger coarse-grained granule ripples, Bagnold reasoned that the commonest mechanism was by the elastic rebound of sand grains upon the local substrate.

The basis of my picture is that the direct motive power which acts on the grains and arranges them into ripples is not the wind but a continued low-angle bombardment by flying grains which are themselves set in motion by the wind…Momentum must therefore be continually extracted from the air; which is another way of saying that the air suffers a continuous resistance while the sand-cloud is being maintained by it…[and] that rebound from impact on pebbles or large sand grains on the ground surface supplies the upward force required to keep sand grains aloft. Laboratory experiments seem to show that this hypothetical bouncing motion of desert sand grains is possible and indeed probable.

In fact, Bagnold missed out on William Chepil's later experiments in the 1950s and his conclusion that the role of a lift force acting on sand grains (as on an aerofoil) could also be an important parameter.

First wind tunnel experiments

Bagnold had already made the acquaintance of C.M. White, later to become Professor of Hydraulics at Imperial College. Between 1933 and 1939 at White's revitalized Hydraulics Laboratory, with R.V. Burns, F.C. Colebrook, E.F. Gibbs, R.P. Pendennis-Wallis and Bagnold himself, fundamental research also took place into applied topics such as wave-pressure, spillway design, pipe and channel friction, cavitation, drag force and flood prediction. Bagnold contributed to some of these hydraulic projects as well as to his own wind-blown sediment transport studies. In laboratory space provided for him at the college by White he began meticulous experimental study of wind-blown sediment transport to support his field observations.

To answer some of the problems raised in the speculative 1935 paper he began construction of a 10 m long experimental wind tunnel within which, using collectors and pressure-measuring apparatus, he could accurately measure both the flow characteristics of pure moving air and of the same magnitude of flow over carefully prepared loose bedstocks of known grain sizes (measured by sieving). The wooden tunnel was constructed with glass sections for viewing purposes, each divided using flexible seals and suspended from gimbals that allowed the weight of sediment being transported into each section over a given time to be calculated. At the tunnel entrance a reservoir of sand was arranged so as to give a steady supply without the blowing air removing sand down to the bare tunnel floor. Air flow was provided by a powerful electric fan sucking air through the tunnel at various speeds under closely controlled conditions. The speed of the wind at various heights above the static or moving sand was computed by pressure measurements made by a fine-bore pitot-tube connected to a manometer (pressure gauge).

Flow measurements at various rates of mean air flow were first made over static beds of carefully levelled sand made moist and immovable by an aqueous 'atomizer' spray. The sand beds were then allowed to dry out, and measurements made to determine the flow threshold for grain movement and then again during increased rates of mean

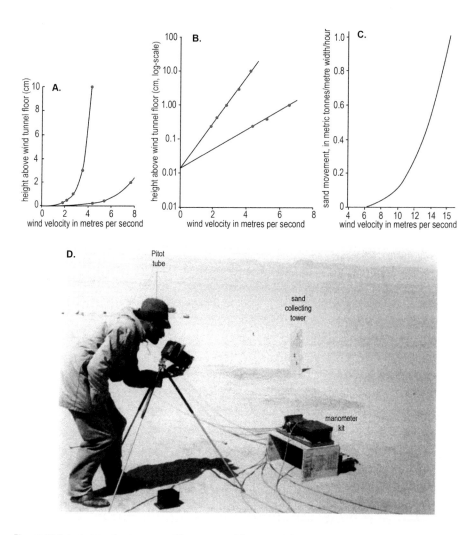

Figure 11.3 A, B. Wind velocity profiles measured by Bagnold using a pitot-tube manometer in his Imperial College wind tunnel in the late 1930s. A. Plot shows the nature of the upward change in velocity defining the wind-flow boundary layer. B. Plot with a logarithmic height scale that conformed with Ludwig Prandtl's 'law-of-the-wall' for turbulent boundary layers. C. Curve to illustrate Bagnold's sediment transport equation relating the magnitude of sand transport by the wind by a cubic function of the shear velocity. It emphasizes the importance of extreme events in sculpturing desert landscapes. D. Bagnold undertaking an 'outdoor physics' experiment by measuring bedload transport during a sandstorm near Gilf Kebir, SW Egypt in 1938. He is photographing his multi-manometer kit whose spaghetti-like connections are leading away to the left towards a pitot-tube (pressure-measuring probe) whose tall tip may be faintly seen directly above his hat. Centre right is the sand collecting column used to measure the transported mass of saltating grains in bedload at various heights in the boundary layer. Image courtesy of University of Arizona Press.

air flow with active sediment transport (Fig. 11.3). The experimental results showed that the rate of increase in flow velocity over static sand beds decreased upwards in proportion to the logarithm of height according to the 'law-of-the-wall' established by Prandtl, with each velocity 'ray' focussing on a common height proportional to the mean diameter of the sand – the so-called roughness height.

Measured velocity gradients enabled computation of the mean shear stress exerted by the airflows on the sand substrate and the determination of threshold values of this shearing stress for sediment transport. By way of contrast, flow over dry, loose sand beds above the threshold shear stress for motion took the form of a definite dense grain carpet several grain diameters high, accompanied by the passage of ripple-like bedforms.

Photographic images showed grains suddenly and steeply rising and then more gradually descending back into the carpet, the saltation motions (Latin: *saltare*, jumping or dancing) originally described and named by G.K. Gilbert of the United States Geological Survey in 1914. Velocity measurements in these experimental runs also obeyed the law-of-the-wall in the outer airflow, but the velocity rays measured within the carpets showed a reduction in flow velocity consistent with the transfer of fluid momentum from the air flow to the sediment flow.

The ability to calculate sediment transport rates for windblown sands has many applications, not least from the point of view of natural hazards to mankind and damage to communications, pipelines and agriculture. To this end, Bagnold made a simple balance of forces for the effect of bedload on wind flow such that the available wind power for transport was proportional to the cube of the flow shear 'velocity' – a measure of the velocity gradient (and applied stress) derived from his velocity ray profiles. This discovery of a markedly non-linear increase of transport capacity was confirmed by sand carpet mass measurements over the adjustable length sections of the experimental tunnel. Later field measurements (Fig. 11.3D) of velocity profiles and transport rates using ingenious bits of portable apparatus carried out in Egypt during 1938 (he spent two months waiting for a decent sandstorm) were entirely consistent with the experimental results of the previous year.

War (again), wind and water

Early during the Second World War, Bagnold was recalled to the army as full Colonel and found himself on active service once more, founding, resourcing and commanding out of Cairo the Long-Range Desert Group in their exhausting patrols westwards. This irregular scouting force, initially of New Zealanders, also acted piratically by agreement with General Archibald Wavell, 'one-eyed Wavell', as Bagnold referred to him in conversation, supporting General Richard O'Connor's spectacularly successful campaign against Mussolini's much larger forces in North Africa. Also in 1941, his book *The Physics of Blown Sand and Desert Dunes* was published back in England and the USA. It became a classic of quantitative sedimentology and geomorphology, still in print and widely admired 80 years later.

During the war his Imperial College colleagues undertook studies concerned with the design of pneumatic breakwaters for Mulberry Harbour, and similar projects. Some of this work used results from Bagnold's results on the pressures exerted by water waves on jetties and harbour walls. In 1944 Bagnold's scientific achievements were recognized when he became the only serving officer in the British armed forces ever to be elected to fellowship of the Royal Society of London.

At the end of the war, leaving the army for the second and last time, on this occasion with the rank of Honorary Brigadier, he continued his research into the sediment transport systems of the earth's surface with fundamental contributions to the problem of the discipline that he was to name 'loose-boundary hydraulics'. His work on wind-blown transport in both laboratory and field had emphasized the impossibility of directly measuring the forces acting upon grains moving against the force of gravity. In 1954 he published a fundamental paper that described how

Figure 11.4 The consequences of the 'Bagnold Number' effect in granular/fluid mixtures under shear is illustrated in this 10 cm wide, finely crystalline, dolerite dyke (grey colour) with its prominent large black pyroxene crystals ('phenocrysts') from the island of Rhum, Scotland. The sharp, 'chilled' intrusive contacts between the dyke and its olivine-rich (weathered light brown) host (a dunite) are followed by a marked concentration of the larger pyroxene crystals towards the dyke centre. During intrusion these were forced inwards from the dyke's shearing boundary layers (see Fig. 9.8) by crystal-induced pressure gradients until preserved as the magma cooled rapidly below its solidus and motion ceased.

he overcame this stricture by means of an elegant experiment on the behaviour of neutrally buoyant, solid (hard lead-wax), spherical grains in the water-filled annular space between a coaxial rotating outer drum and a stationary inner one. The wall of the latter, in contact with the rotating fluid–grain mixtures, was deformable and able to transmit the resulting pressure and torque caused by both fluid shear and grain-to-grain collisions to a manometer measuring system.

By utilizing pure water and steadily increasing the concentration of grains and varying the fluid shear rate in the rotating drum, he was able to identify the chief variables that controlled the behaviour of the grain/fluid system. The dimensionless number formed by the ratio of these several variables that he was able to derive and compute has since become known as the Bagnold Number. Its value determines whether a grain/fluid system is 'viscous-controlled' (with fluid viscosity dominating over the effects of solid-to-solid grain collisions) or 'inertia-controlled' (with grain collisions dominating). The criterion is flexible and general so that it can be used to describe the flow of blood and platelets in veins and arteries, the flow and interaction between liquid magma and crystals (Fig. 11.4) and the bedload transport of quartz sand in rivers – Ralph Bagnold was a worthy and ingenious successor to Osborne Reynolds.

Rivers as transporting machines

Bagnold had a late outpouring of creativity with the United States Geological Survey as a visiting scientist in the late 1950s and 60s, courtesy of an invitation by the leading US geomorphologist of his generation, Luna Leopold, son of Aldo Leopold, acclaimed pioneering ecologist and environmentalist. Bagnold wrote of his friend:

> Here was a man of strong character and great practical experience, a nationwide authority in his field [of fluvial geomorphology], and with the backing of a far-sighted director. We agreed that we ought to 'stir the pool of complacent tradition with the stick of reality'. He suggested that with his practical help and influence, I should try to find the stick – an explanation of how the laws of nature operate to make rivers behave as they do in moulding their channels according to their individual flow conditions. All of the US River records would be available to me.

He worked for several years on the transport of sediment by water flows using old and new data provided by the hydrologists of the USGS, mainly at his home in Edenbridge, Kent, with annual visits to the US with his wife, combining work with enjoyable travels in the deserts of the SW USA. The result of this fruitful collaboration was a severely pruned and revamped version of his overlong and somewhat discursive 1956 paper 'On the flow of cohesionless grains in fluids', published in the *Transactions of the Royal Society of London* – notable not only for its length but also for the brevity of its two-author bibliography: Albert Einstein, Osborne Reynolds!

The new work came out in 1966 under the imprint of a *Professional Paper of the United States Geological Survey* as 'An Approach to the Sediment Transport Problem from General Physics'. In it Bagnold developed his mature concept of flowing water as a transporting machine that works with a certain efficiency by applying its available power gained from the conversion of fluid potential energy to kinetic energy under

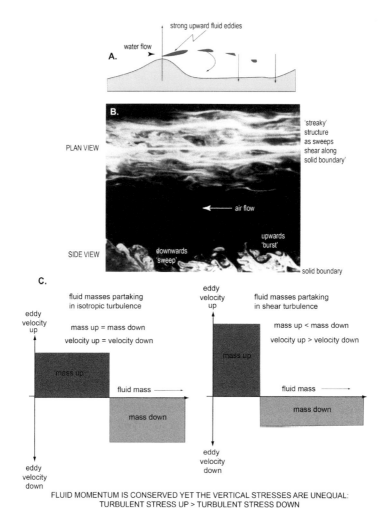

Figure 11.5 Bagnold's entirely original (and widely misunderstood) 1966 mechanism for sediment suspension by fluid turbulence. A. His sketch of L. Prandtl's photograph of water flow separation eddies over a wave-like form showing the asymmetry of turbulence. B. A modern smoke visualization by R.E. Falco (1977) of asymmetric eddies in wind flow above a rigid lower boundary due to smaller upward 'bursts' and larger downwards 'sweeps' of fluid. C. Bagnold's reasoning concerning the inequality of vertical turbulent stresses – the excess of up-stress providing the motive power to hold up a suspended mass of sediment in wind or water.

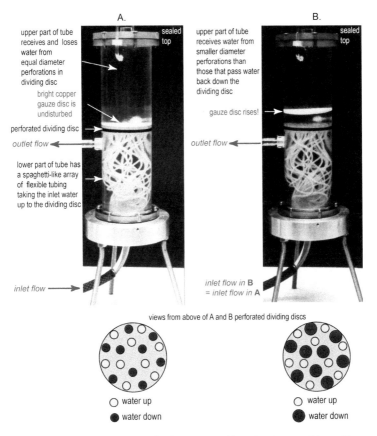

Figure 11.6 Two experimental 'Bagnold Tubes' constructed by John Mott and his colleagues in the well-appointed workshops of the Department of Earth Sciences, University of Leeds in the mid-1990s. From an original concept and design by Ralph Bagnold. When in action they demonstrate the net upwards stress needed to support a suspended mass, in this case a thin disc of copper gauze. A spectacular teaching experiment!

gravity. He made major breakthroughs in his physical attempts to define both power availability and efficiency, testing his overall predictive equations with field and experimental data.

A particularly novel aspect to his approach was a highly original dynamic slant to the treatment of suspended load, including expressions for the energetics of this process that had fascinated him since his desert days of the 1930s (Figs 11.5, 11.6). This was based upon his view that turbulent energy production in boundary layers was anisotropic, with upward momentum generated by smaller but faster jets of fluid than the sluggish but larger masses of returning downward fluid. Although momentum is necessarily conserved by this arrangement, the imbalance in up- and down-velocities led to a net upward stress that would enable a suspended load to be kept up in a flow above a carpet of bedload. His views were based on experimental fluid mechanical measurements by S. Irmay and by a theoretical treatment by A.A. Townsend, a leading light on the fluid mechanics of turbulence. Later calculations and experiments by the present author and colleagues at UEA Norwich and by Sean Bennett and colleagues at SUNY Buffalo entirely supported his approach.

One major problem encountered in his workings through the USGS field and experimental data was the highly varied nature of experimental conditions recorded. Bagnold's solution was enabled by careful experimental work carried out by Garnett Williams of the USGS on the hydraulic effects of flume wall drag and flow depth on the magnitude of available flow power, flow efficiency. The end-result was his last paper, *Transport of solids by natural water flow: evidence for a worldwide correlation*, published in 1986 when he was 90 years old – a fitting end to an exceptionally long and original scientific career.

PART 5

Climatic Stuff: Settings

Milutin Milanković (1879–1958); Harold Clayton Urey (1893–1981)

The nature of past climate has long fascinated humankind. Early nineteenth-century geologists realized that thick and extensive deposits of mineral salts in certain strata must have formed by the evaporation of seawater to form brines in subtropical lagoons and lakes. Since these deposits were now located in higher latitudes, often in temperate climatic regimes, the implied latitudinal change required a serious effort of explanation that took over a century to achieve – in fact, not until Wegener's continental drift theory were such changes finally confirmed by palaeomagnetic studies.

Geologically more recent and severe climatic oscillations were recognized when Louis Agassiz's 1840s field evidence for the existence of former ice age advances and retreats, gathered from around the outer Swiss Alpine valleys in the 1840s, fundamentally challenged the existing order. Eminent British geologists like Lyell and William Buckland were ready converts to his field demonstrations of glacial landforms and their witness to former glacial action in the mountainous districts of Scotland (indeed, on Lyell's home doorstep) and Ireland on his celebrated visit from Switzerland. However, resistance to such field evidence by the 'old guard' of the Geological Society of London prevailed and the glacial cause was abandoned for a generation. By the 1860s it had been firmly adopted by the younger and more perceptive of field geologists. It was then that an explanation for the causes of periodic climate shifts between temperate and sub-polar conditions became pressing.

This was a challenge that James Croll in Glasgow, following earlier suggestions by Lyell and John Herschel, took up by building on the planetary orbital theories of Joseph Adhémar and Urbain Léverrier. Croll's predictions of alternating cold and warm climatic interludes in the Pleistocene due to periodic orbital controls were hampered by the lack of a geological timescale. Encouraged by Alfred Wegener and Wladimir Köppen, the young Serbian engineer Milutin Milanković took up Croll's mantle in 1913. With his advanced mathematical knowledge of how to determine radiative heat transfer into the upper planetary atmosphere, he began to compute his way through a whole decade of his life to arrive at a time series of changing

Pleistocene climate that was incorporated into Wegener and Köppen's *Climates of the Past* in 1924.

Harold Urey made huge contributions to the science of palaeoclimate, first through his discovery of heavy hydrogen, a stable isotope that he christened deuterium (Greek: *deuteros*, second, re its two-particle nucleus) and for which he gained the Nobel Prize for chemistry in 1934. Second, he developed understanding of the role of the stable isotopes of oxygen and carbon in their reactions with a small number of other elements involved in the great water, oxygen and carbon cycles of atmosphere, hydrosphere and biosphere. From his fundamental work on the reaction mechanisms of isotope fractionation he also devised a method for determining the ancient temperature of seawater from oxygen isotopes, which in turn, 15 years or so later, enabled the discovery of minute isotopic fluctuations of the oceans in response to changing glacial and interglacial conditions. These led to the confirmation of the orbital control on climate. He also pioneered the concept of the environmental cycling of the element carbon, and encouraged his doctoral student Stanley Miller's experimental insight into the likely processes that might have created the first life forms from amino acids produced in the oxygen-deficient atmosphere of early Earth.

12

Milutin Milanković (1879–1958) *Serbian applied mathematician and palaeoclimatologist*

Milanković some time in the early 1920s. Wikipedia Commons.

Following on from James Croll's pioneering work of the 1860s to 70s he derived and computed the magnitude of insolation by incoming solar radiation to determine palaeotemperatures at various latitudes over the four final ice ages known at the time, confirming their control by cyclical orbital motions (1920, 1924, 1938). Such climatic cyclicity is often referred to as 'Milanković cyclicity' but in the light of Croll's work (he did not 'labour in vain') it is perhaps more justified to refer to it as 'Milanković–Croll cyclicity'.

A Serbian polymath

Just a few months after Alfred Wegener's historic Frankfurt-am-Main lecture on continental drift in January 1912, a Serbian civil engineer with mathematical interests submitted his first scientific paper *Contribution to the mathematical theory of climate*. As with Wegener's offering, it marked an historic break with nineteenth-century science. It had arisen from his leisure time reading around the newly emerging sciences of meteorology and climatology. The engineer was Milutin Milanković and, rather like Wegener with drift, he had stepped outside of his immediate comfort zone to tackle an interesting and topical problem – the origins of ice ages. As he wrote later, it seemed to him that:

> most of meteorology is nothing but a collection of innumerable empirical findings, mainly numerical data, with traces of physics used to explain some of them... Mathematics was even less applied, nothing more than elementary calculus... Advanced mathematics had no role in that science...

This was a man who could calculate the mathematics behind the passage of incoming solar radiation. His reading had also made him familiar with previous efforts to understand past climates through the movement of the planets in their own changing rotations and in their orbital passages around the Sun. This came chiefly through the work of James Croll, whose extraordinary life and work we expand on below. Milanković decided to follow in Croll's footsteps and to attempt to bring the full force of advanced mathematical geometry and general physics to bear on the problem of the ice ages. His first published work described the present climate on Earth and how the incident Sun's rays that comprise solar insolation determine the temperature at the top of the atmosphere. It was followed by research on past climates, with his chief results first presented to the world at large through Klöppen and Wegener's book, *Climates of the Past*, discussed in Chapter 5 and further herein.

James Croll – who he?

Croll was Scottish, from a long-established Perthshire crofting family in the parish of Cargill. Like so many others they were displaced (locally) in clearances by landowner Lord Willoughby (16th Baron Willoughby de Broke) in the 1820s. His father, David, was a stonemason, an observant Congregationalist and a kind, serious and thoughtful man. His mother, Janet Ellis, a practical woman, was originally from Elgin in Morayshire. The couple were married in 1818 and had four sons. James was the middle child born in 1821, their eldest son dying aged 10 and the youngest as an infant. James himself was a sickly child – bad headaches kept him away from local schooling in his young years; times when he helped his mother out on the Wolfhill family croft while his father pursued his itinerant craft. Like many unwell children he became an avid reader in his spare time, this tending to the scientific and philosophic side as he grew older and began to be attracted by the logical nature of such subjects.

He was taught badly when he first re-joined school, but eventually when he was thirteen a sympathetic teacher in an adjoining parish gave him the encouragement that he needed. But this was only for a few months before he was taken from school permanently to give his mother help in the home and on the croft. From the age of eleven his private hobbies were aided by his discovery of the popular science *Penny Magazine* and the sterner *Christian Philosopher*. He writes in his autobiography that:

> I shortly afterwards procured one or two other books on physical sciences, among which was Joyce's famous scientific dialogues [up to 1805: 'Intended for the Instruction and Entertainment of Young People: in which the First Principles of Natural and Experimental Philosophy are Fully Explained']. At first I became bewildered, but soon the beauty and simplicity of the conceptions filled me with delight and astonishment, and I began then in earnest to study the matter.

In light of his later scientific exploits, it comes as a ray of understanding to recognize Croll's affinities with the pioneers examined in this book when we also read:

> I may mention that, even at the commencement of my studies, it was not the facts and details of the physical sciences which riveted my attention, but the laws or principles which they were intended to illustrate.

Croll goes on to itemize the kinds of scientific laws and principles he studied in this manner over the next four years or so: physical astronomy, the laws of motion, mechanics, pneumatics, hydrostatics, light, heat, electricity and magnetism. A pretty precocious set of attainments!

Subsequently, during a difficult life, Croll progressed by entirely private study from adolescent amateur scientist to a published metaphysical philosopher of the Christian kind. By 1859 the startlingly varied day jobs necessary to provide for him and his beloved, caring and doubtless long-suffering wife, Isabella, over the past 20 years included millwright, carpenter, tea retailer, electrical engineer, hotelier, insurance salesman and journalist. Now approaching middle years, after a major health scare, he had applied for the advertized post of janitor to Anderson's University in Glasgow, along with 60 or so hopeful others. The University (its descendant is the University of Strathclyde) was unique in Britain, for it was founded in the late eighteenth century as a place of learning in the 'useful' applied sciences. Croll got the job, the provender of employment being Walter Crum, chair of the Anderson's directors and himself an industrial textile chemist of some repute. It seems that Crum had met Croll occasionally whilst on countryside walks and had been impressed with the obvious seriousness of Croll's interests in science and philosophy.

Croll's new job started in the late-autumn college term of 1859 and, though mundane in its duties and hardly well paid, it provided regular and safe employment and, more important to Croll, access to the fine 2000-volume scientific library of the Glasgow Philosophical Society kept in the College. It enabled him to bring to rapid fruition within the next few years original studies on the nature and consequences of long-term climate change and its controls. The post not only gave him time to study, it enabled him to rise through, so to speak, a glass ceiling into the privileged world of serious science. In fact, Crum's son-in-law was the already-eminent physicist William Thomson of the University of Glasgow, later to become the Lord Kelvin whom we met in Chapter 7, destined in time to write one of Croll's many scientific obituaries. By 1864 Croll had mastered the current state of knowledge concerning possible orbital controls of past climate that might have led to the ice ages (Fig. 12.1). He later wrote in his autobiography:

> In the spring of 1864, I turned my attention to this subject; and, without knowing at the time what Herschel and Lyell had written on the matter, it occurred to me that the change in the eccentricity of the earth's orbit might probably be the real cause...I resolved to follow it out. But little did I suspect...that fully twenty years would elapse before I could get out of it.

Croll had established to his own satisfaction that climate had changed in the geological past but that there was no single theory extant to explain such changes. The

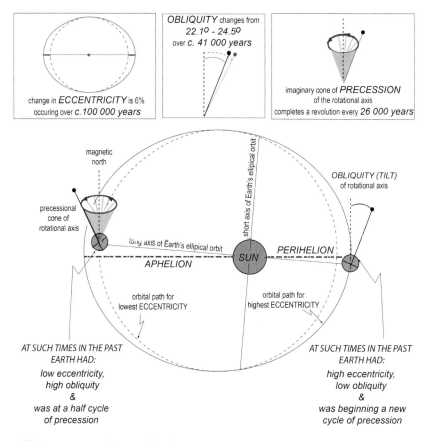

Figure 12.1 Diagrams to illustrate the relevant orbital behaviours of the earth that determine radiative heat transfer to the upper atmosphere during its journeys through space around the sun.

nearest was that presented by Herschel and Lyell – the likelihood of cooler northern hemisphere temperatures during phases of planetary tilt outwards from the ecliptic at the winter solstice, vice versa for phases of tilt inwards. But until Croll's first work was published, Lyell always regarded orbital effects as minor ones in climate control, favouring the changing distribution of land masses over the globe as more important.

Three original papers by Croll appeared in the *Philosophical Magazine* between 1864 and 1867 concerning the control of climate by orbital perturbations. His 1864 contribution begins:

> No fact in geological science is better established than that in former periods of our earth's history great changes of climate … must have taken place … yet there is the greatest diversity of opinion regarding their cause and origin.

He quickly surveyed the published reasons for this diversity of opinion, dismissing each in turn according to various authors and to his own rationality – the role of internal heat, wholesale change in the rotation axis, passage of the planet through hot and cold areas of space, different geographic distributions of land and sea (the latter from Lyell). He then states his guiding opinion that 'the true cosmical cause must be sought for in the relations of our earth to the sun.' In conclusion:

> There are two causes affecting the position of the earth in relation to the sun, which must, to a very large extent, influence the earth's climate: *viz.*, the precession of the equinoxes and the change in the eccentricity of the earth's orbit. If we duly examine the combined influence of these two causes, we shall find that the northern and southern portions of the globe are subject to an excessively slow secular change of climate, consisting in a slow periodic change of alternate warmer and colder cycles.

Croll had realized that, in addition to the effect arising from precession, the upper limit to the earth's orbital eccentricity as determined by Urbain Leverrier in 1858, around six percent, was large enough to have a significant effect on the seasonal radiative heat received by earth's hemispheres. It would significantly increase northern hemisphere winter distance and therefore reduction of solar radiation at aphelion (Fig. 12.1). Severer northern winters would result. With the assumption that winter snowfall would significantly increase and that warmer summer rains and greater insolation at perihelion would be unable to melt all the extra snow, he assumed that periods of high eccentricity would tend to be glacial.

He also pointed out that snowfields possessed their own dynamics by reflecting incoming solar radiation and therefore amplifying the loss of winter heating capacity. This was the first published statement of what we now know as 'positive feedback'. He added that a similar role could be played by the diversion of warm ocean currents from the mid-latitudes by ice accumulation and weakening of the trade wind systems that drove them – notably the North Atlantic Gulf Stream.

Then he played his trump card. When rotational precession during aphelion at times of high eccentricity pointed the northern hemisphere away from the sun then radiation reduction would be multiplied, as he determined, up to about a fifth. Therefore, severe glaciation would occur when eccentricity and precessional tendencies were acting together to attenuate incoming solar radiation. Southern winter climatic frigidity and glacials would alternate with those of the northern hemisphere within the precessional cycle every 11 ky. With low eccentricity, when the tendencies in both hemispheres were in opposition, there was little effect in a more or less circular earth orbit. Croll ended his discussion with a paragraph that would echo down the ages, for it concerned the chronology of Leverrier's eccentricity changes and therefore of glacial periods:

> As yet no calculations have been made regarding the time when the eccentricity was at a maximum, or of the time required to pass from the maximum to the minimum state of eccentricity. In the Annales of the Paris Observatory…there is a Table giving .0473 for the eccentricity at 100 000 years before the year 1800, and .0189 for the eccentricity 100 000 years after 1800. There are subordinate maxima and minima in that interval of 200 000 years; but the principal maximum I have been informed does not fall within that period. We may therefore safely conclude that it is considerably more than 100 000 years since the glacial epoch…

This flourishing finish gave the world its first independent estimate for the chronology of the last Ice Age. There has been no suggestion as to who it was who informed Croll of the age of the principal maximum – it may have been Leverrier himself, by letter, or from a certain Mr Stone at Greenwich Observatory, who, three years later, contributed to Croll's second paper (see below).

Croll could not have chosen a more propitious time to make his theory known. John Imbrie and Kathleen Palmer Imbrie, in their pioneering account of the history of orbital theory, *Ice Ages: Solving the Mystery* (1979), write:

> By 1864, William Buckland and Charles Lyell had been won over by Agassiz, the ice-age concept had been almost universally accepted, and geologists were eager to find an explanation for the ice-age cycle.

In 1867 Croll published two further papers that explored Leverrier's eccentricity data, adding numerous new interpolations by Croll himself and by Stone of Greenwich Observatory. The result was the first astronomical time series ever published (Fig. 12.2). The new data proved beyond doubt that orbital eccentricity changes were cyclical, with peaks of highest eccentricity separated by troughs of low eccentricity over tens

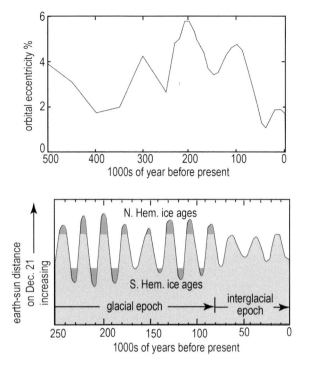

Figure 12.2 Croll's time-series for variation through geological time of A. orbital eccentricity and B. the Earth–Sun distance at northern hemisphere winter solstice. The shaded peaks in B are times when Croll believed that ice ages would be triggered by the onset of the critical distance needed to produce successive colder winters where enough seasonal ice could accumulate, so leading to glacial expansion and ice ages. From Croll, 1867.

to hundreds of thousands of years. The time series also featured estimates made by Croll of the number of days that the length of the northern hemisphere winter in aphelion exceeds that of the summer and the corresponding temperatures by which the midwinter temperature was lowered.

Towards the end of his paper Croll returned to the problem of chronology and those high peaks of eccentricity between 100 and 700 ka and asked whether in fact they would place the last Ice Age '… too far back'. Could it really be 100 ky since the end of the period of boulder-clay [glacial deposits] formation? He also wondered whether reported Miocene glacial deposits might be older, leaving the 240 ka to 80 ka eccentricity highs causing the more recent glacial epochs.

The core of Croll's second 1867 paper added yet another orbital parameter to the mix of precession and eccentricity. This concerned Pierre-Simon Laplace's determination that the tilt of the rotational axis (obliquity) could change relative to the ecliptic plane by around ±5% – its present value is 23.5 degrees. Croll pointed out that such changes would have relatively little effect on low and middle latitudes at periods of high eccentricity and precessional alignment, but they would be important at high latitudes. Unfortunately, nothing was known at the time concerning the chronology of such changes and so the effect could not be added to the known effects of eccentricity and precession.

Re-enter Milutin Milanković

By 1913 Milanković was sure of his future course of action: the construction of an overreaching mathematical theory of climate, delimiting the chief climatic zones as thermal regimes responding to the quantity of incoming solar radiation as controlled by orbital fluctuations. His ambition was not only to accurately compute modern climate but also to peer into the geological past and search for the objective reasons for what were once regarded as mere climatic aberrations – to shine a bright and mathematical light onto their true origins. In a 1913 contribution, *Distribution of the sun's radiation on the earth's surface*, he writes:

> ...such a theory would enable us to go beyond the range of direct observations, not only in space, but also in time... It would allow reconstruction of the Earth's climate, and also its predictions, as well as give us the first reliable data about the climate conditions on other planets.

It was an ambitious and lengthy project for those times before computers, on a par with Leverrier's vast calculations of 60 years previously. His preliminary discussion of the necessary contents of such a scheme came in 1914 with a paper titled *About the issue of the astronomical theory of ice ages*. However, his work and his life plans were to be brutally changed that year by the outbreak of World War I, triggered by his native Serbia's strained relationship with the Austro-Hungarian empire. Unaware of the outbreak of hostilities (he was on his honeymoon in his home village in Austro-Hungary at the time) he was arrested as a Serbian citizen and imprisoned. Here he describes his initial reactions to imprisonment – what might normally be viewed as an awful situation he saw was an opportunity and became determined not to let it change his life:

> The heavy iron door closed behind me...I sat on my bed, looked around the room and started to take in my new social circumstances... In my hand luggage which I brought with me were my already printed or only started works on my cosmic problem; there was even some blank paper. I looked over my works, took my faithful ink pen and started to write and calculate... When after midnight I looked around in the room, I needed some time to realize where I was. The small room seemed to me like an accommodation for one night during my voyage in the Universe.

Meanwhile, his resourceful wife had gone to Vienna and with the help of influential friends in the university was able to arrange his transfer to Budapest with a right to continue his research, and where he was welcomed by a fellow mathematician and given access to the Hungarian Academy of Sciences library. He spent the rest of the war developing his astronomical control of climate ideas, presenting results of the likely climatic conditions on the planets Mars, Venus and Mercury as well as our own Moon. After the war and a move back to the University of Belgrade with his family in 1920 he arranged for the French publication of his work to date, the *Mathematical Theory of Heat Phenomena Produced by Solar Radiation*. It marked the beginning of modern climate modelling studies, for he had reduced six elliptical elements of spherical geometry down to two vectors that determined the mechanics of planetary movements.

True cyclicity of ice ages

It was the magnitude of changes in incoming radiation felt at the *top* of the atmosphere that had to be calculated by Milanković, not the far more complex energy exchanges and temperature behaviour controls on meteorological climate induced in the atmosphere itself. It was impossible 100 years ago to compute how the atmosphere and the oceans actually react to incoming and outgoing radiative changes – our modern-day climate change.

It is here that we turn once more to the steering arms of Wladimir Köppen whose intellectual and family relationships with Alfred Wegener have already been featured (Chapter 5). Both men had read and admired Milanković's 1920 book, and

Figure 12.3 Late-summer snowfall surviving on the north-facing slopes of the Helvetic Alps above Martigny, Switzerland in 1975. Foreground shows structural geologist John Ramsay (E.M. Anderson's modern successor) with his mapping group of Leeds undergraduates – John was renowned for his close interest in field supervision.

in correspondence had persuaded him that it was not winter, but summer insolation that was the chief factor for ice age conditions by glacial expansion to become established. This meant that previous winters' snow accumulations needed not only to be abundant, but they also needed to be preserved through successive summer heat for hundreds or thousands of years to enable ice build-up (Fig. 12.3). Incoming radiation creating this 'summer sunshine' needed to be calculated.

Late in his long life, Vladimir Köppen gave a concise analysis of Milanković's computational procedures in his 1938 revision notes for the second edition of *Climates of the geological past*:

> The fluctuations of solar insolation reaching the boundary of the atmosphere depend on the changes of three parameters: ε the obliquity of the ecliptic, *e* the eccentricity of the earth's orbit, and π the length of the perihelion. The latter two parameters invariably appear related to each other in the term *e* sin π. These changes were calculated from Newton's law of gravity according to the configurations and masses of the planets and the sun, the result therefore depends on data pertaining to these masses, most of which could only be determined in recent times. It is a mistake to believe that the equations were extrapolated from observed changes.

He went on to raise a fundamental point with respect to ancient climates and their relations to continental drift, one which would remain unsolved until the 1970s. He writes [with my italics]:

> These fluctuations of ε, *e* and π and the resulting fluctuations of the earth's insolation certainly did not occur only during Quaternary, instead, *they were continuously there throughout all times*, however, the climate change differed in space and time, depending on the changing location of a site relative to the pole and the equator, and to water and land.

In a letter of the early 1920s, Köppen suggested that Milanković expand the extent of his existing 130 000-year insolation curves for summer sunshine to 600 000 years. In enthusiastic response to this recognition by the world's most pre-eminent climatologist, Milanković spent three months computing a graph of insolation for latitudes of 55°, 60° and 65° north – the latitudes on earth he thought most sensitive to the change of thermal balance that might allow 'global' ice ages to establish themselves. This epic solar curve (Fig. 12.4) was featured in fine style as a pullout in Köppen and Wegener's book (with Milanković's permission) with the fullest acknowledgements, immediately establishing his work (and the book itself) as a climatological and geological classic – the science of palaeoclimatology had arrived.

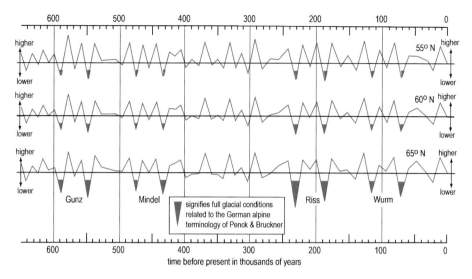

Figure 12.4 Comparative (higher to lower) solar insolation curves for the summer half-year ('summer sunlight') in higher latitudes (55, 60, 65 °N) during the last 600 000 years – middle/upper Pleistocene times – the original Milanković time series redrawn from Köppen and Wegener (1924). Gunz, Mindel, etc. are the supposed correlations with the German alpine glaciations proposed in classic work by Penck and Brückner a decade or so earlier. The Würm event should include the final dip in the insolation curve – the last Pleistocene glaciation; see Fig. 12.5.

Post-1924 recalculations by Milanković and Vojislav Mišcović

Fast-forwarding to 1938, Köppen made further revealing comments and insights into the status and provenance of Milanković's work in the notes attached to a reprinting of *Climates of the Past*. He refers to the fact that there had been important confirmations of the Alpine results of Penck and Brückner over the past 14 years, with field evidence for distinct ancient glaciations across Germany that must coincide with the latest four recognized in their classic work – Günz, Mindel, Riss and Würm (see Fig. 12.4). These had been put into Milanković's time series by himself and Wegener in 1924. He added the following account of post-1924 events affecting Milanković's insolation calculations:

> The – very complicated – calculation of the fluctuations of ε, *e* and π on this basis has been carried independently from each other by Leverrier and Stockwell, in both cases with subsequent corrections due to changes in basic parameters. Pilgrim made the numeric evaluation in 1904, based on Stockwell's equations, and Milanković used these values in our book. However, as it showed that some of Pilgrim's figures in part had calculation

errors, in part had been obtained by too far-reaching interpolations, Miscovic, Professor of astronomy in Belgrade, repeated the calculations for the purpose of corroboration, this time applying the equations of Leverrier and the latest mass determinations. Hence Milanković determined the radiation parameters for various geographical latitudes anew...Our Figure 3 shows... the new calculation for a latitude of 65°. In the main, it is a very appreciated independent corroboration of the representation from 1924, but it shows that last jag [cusp] corresponding to Würm III 22 000 years ago, much stronger, like a real glaciation period.

These new curves of Milanković are shown in Figure 12.5. Later developments in the chronology of the ice ages by radioactive and magnetic dating, isotope geochemistry and orbital modelling would finally confirm his life's work in 1976 (Chapter 18).

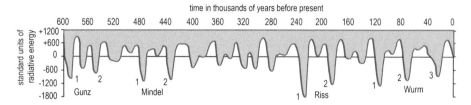

Figure 12.5 Milanković's 1938 recalculated summer half-year radiation received at 45 °N. The curve includes not only the standard forcing by all three orbital controls but also by calculated changes in the earth's reflective capability, its albedo. The 23 000-year precession cycle stands out clearly at this latitude. The last glaciation, Würm 3, was the first ice age experienced by *Homo sapiens* and which we now know to have ended progressively from around 20 000 years before present.

13

Harold Clayton Urey (1893–1981)
USA geochemist and planetologist

Urey pictured in 1934, aged 41, at the time of his Nobel prize. Wikimedia Commons.

He discovered deuterium, the stable isotope of hydrogen, in 1931 and was awarded the Nobel Prize for chemistry in 1934. In the 1950s he linked oxygen stable isotope ratios with past climatic and oceanographic temperatures and defined the basic principles of geological-scale carbon cycling via a buffering reaction in the chemical weathering of calcium silicate minerals by carbon dioxide. He inspired his graduate student, Stanley Miller, to experimentally produce amino acids in a laboratory-controlled, oxygen-free environment that mimicked the likely conditions (according to Urey) on the young Earth (1953).

Elemental fossils

Continuing with the theme of 'modern alchemy' that began with Norman Bowen's experiments on magmatic crystallization, we turn back to the fundamental discovery by Frederick Soddy in 1913 of what he and a classically educated family friend – the pioneer Scottish medic and novelist, Margaret Todd – named as 'isotopes'. The word is nicely coined from Greek roots: *isos*; same and *tope*; place, for these are less-common variants of 'normal' elements but with the same atomic number (the number of protons in the nucleus) and therefore the 'same place' in the periodic table of elements. Their different atomic make-up was later found by x-ray spectroscopy to be due to a mass difference caused by different numbers of neutrons in their elemental nuclei (Fig. 13.1)

As we saw in Chapter 7, the first discovered isotopes were unstably radioactive (as radium and uranium) or the stable products of their radioactive decay (as lead). It was gradually realized that there were many other natural examples that were stable, their increased atomic weights giving them slightly different physical chemical properties, in particular their behaviour during mineral-forming chemical reactions, a topic that Urey was to make his own. Such stable isotopes were formed during the early evolution of planetary nuclei – they are elemental 'fossils' – endlessly recycled amongst minerals, natural waters and the atmosphere. Their ability to record conditions of mineral crystallization over geological time established stable isotope geochemistry as a key player in the elucidation of ancient climates from analyses of organic and inorganic mineral precipitates. Much of the pioneer work in this field was accomplished by Urey, his associates and doctoral students.

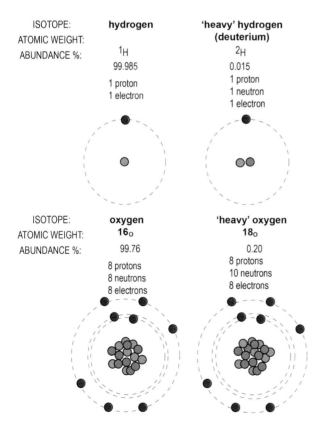

ISOTOPE:	**hydrogen**	**'heavy' hydrogen (deuterium)**
ATOMIC WEIGHT:	^1H	^2H
ABUNDANCE %:	99.985	0.015
	1 proton	1 proton
	1 electron	1 neutron
		1 electron

ISOTOPE:	**oxygen**	**'heavy' oxygen**
ATOMIC WEIGHT:	^{16}O	^{18}O
ABUNDANCE %:	99.76	0.20
	8 protons	8 protons
	8 neutrons	10 neutrons
	8 electrons	8 electrons

Figure 13.1 Simple diagrams to illustrate the various atomic constituents of the stable isotopes of hydrogen and oxygen.

The life

He was the eldest of three children born in Walkerton, Indiana, USA in 1893 to Samuel Clayton Urey and Cora Rebecca Reinoehl. Both parents had college degrees, Samuel, like his forebears, farming on a small scale while both he and Cora taught in local schools. The parents were strict Christian believers and part-time Sunday school teachers and preachers, belonging to a group that aimed to follow to the letter Christ's example in their own lives. Harold was to be brought up with an independent and principled austerity, moral certainty, tolerance and pacifism (modified in the two world wars) with a philosophy of autonomous personal morality that guided certain life-choices he was to make.

That life began badly when father Samuel died young in 1899 and his mother took the children to live with, first, her in-laws on a farm in Corunna, Indiana and then, after their deaths, to near Kendallville, a town of around 5000 people in northern Indiana. By this time Harold had received only a very basic primary education, but fortunately for him and his siblings his father had left the children a legacy that enabled them all to have a full high-school education till the age of 18.

He excelled in his curriculum at Kendallville High School and was able to graduate in 1911 and take up a place at Earlham College in Richmond, Indiana where he obtained a teaching certificate. The college was, and is, a high-achieving Quaker institution founded in 1845, its name a tribute to Joseph John Gurney of Earlham Hall on the western outskirts of Norwich, England, who supported the college in its early years. Urey's achievements at high school and college had been made possible by his father's legacy, something he never forgot, as he wryly recollected later in life: 'If it hadn't been for that, I'd still be in Indiana, working as an unsuccessful farmer – I just can't see me being a successful farmer.'

After a period teaching in country schools, he followed his mother and siblings up to Montana, where he entered the state university in Missoula in 1914. Three years later, he received a BS degree with a major in Zoology and a second major in Chemistry. In his Willard Gibbs Medal address many years later, Urey spoke warmly of the friendships and inspiration he had received from his professors in chemistry and biology at Montana who, he said, largely determined the direction of a scientific career that spanned these disciplines.

After graduating in 1917, the United States having finally entered World War I, Urey took up a wartime position as a research chemist in Philadelphia to help make explosives. After the war he returned to the University of Montana as an instructor in chemistry for two years, and in 1921 he entered the University of California, Berkeley, for graduate work in chemistry with a minor in physics. Berkeley at this time was a hotbed of activity in the chemical and physical sciences, and as a mature graduate student who had already thought deeply about his subject it took him only two years to get his Ph.D. in the labs of the celebrated physical chemist Gilbert N. Lewis.

Lewis is best known for his discovery of the covalent bond, the concept of electron pairs and of acids and bases (Lewis acids and bases). He was a major influence on Urey's life, research and general philosophy of teaching and instruction – well-known for the almost total freedom he gave doctoral students like Urey and for unselfishly letting them claim their own places in the scientific pantheon. Lewis also successfully contributed to the development of chemical thermodynamics, photochemistry, and isotope separation. Whilst at Berkeley, under Lewis's broad umbrella of interests, Urey himself became increasingly fascinated by the newly emergent field of quantum chemistry.

In August 1923, aged 30, he travelled to Denmark to spend a year carrying out postdoctoral research at Niels Bohr's Institute of Theoretical Physics in Copenhagen. His obituarists (K.P. Cohen et al.) write:

> In the previous year Bohr had been awarded the Nobel Prize in Physics for his discovery that electron energy levels in atoms are not continuous – they come in steps. The atom is said to be quantized and can be explored mathematically using quantum mechanics. Bohr had also shown that an atom's outermost electrons – its valence electrons – are of overwhelming importance in determining its chemical properties. In doing so Bohr invented quantum chemistry – a highly mathematical field – and Urey wanted a piece of this action.

But despite having an enjoyable year in Copenhagen, Urey realized that although his ability in mathematics and thermodynamics was high, it was not high enough for him to become a top-drawer theoretical physicist. He decided that his best chance of scientific success would be in experimental rather than theoretical chemistry. So, returning to the USA in 1924, he became a research associate at John Hopkins University in Baltimore, Maryland, and then in 1929, aged 36, he moved to Columbia University as associate professor of chemistry.

'Useful' stable isotopes

The mass of stable isotopes had been computed by Francis Aston just before World War I by means of his invention of the mass spectrometer, an instrument that was able to separate ionized particles according to their atomic mass during their deflected passages through a magnetic field. Being heavier, the stable isotope fractions in the ionized samples could be imaged and counted as distinctive arrivals along the instrumental tube. Resuming his investigations after the war, Aston designed an improved apparatus of greater sensitivity that enabled him to accurately 'weigh' elemental atoms and to identify scores of stable isotopes.

Like all non-radioactive elements, stable isotopes obey Dalton's rule concerning the indestructibility of matter. Distinguished from their purer cousins by those extra neutrons in their nuclei (Fig. 13.1) they can be imagined as staid, stoical and reluctant relatives who must be persuaded by increased temperature to enter chemical reactions along with their more volatile brethren. In the case of the isotopes of hydrogen and oxygen, the extra neutrons 'weigh down' molecules of water, lengthening the oxygen–hydrogen bonds of the water molecule, thus slowing down their continuous heat-produced molecular vibrations and lessening their intrinsic internal free energy. It is more difficult for such molecules to enter naturally occurring chemical reactions, as in the precipitation of calcium carbonate by inorganic or organic means, or to be changed in state from liquid to gas.

It was while at Columbia that Urey achieved the breakthrough that would lead him on into a career of astonishing breadth and diversity that would encompass the fields of not just nuclear physics and chemical thermodynamics, but also in the wider environmental sciences, biology and cosmology. By 1929 he had worked at the forefront of efforts to explain the role of atomic forces in determining the rates and mechanisms for chemical reactions. It was not an accidental choice that led to his interest in stable isotopes, for he soon recognized that their unique passive properties in chemical reactions might lead to promising discoveries concerning the evolution of oceans and atmospheres. There is an interesting anecdote in his obituary concerning his priorities in relation to the earlier discovery of heavy stable isotopes:

> Urey first learnt of the discovery of the rare isotopes of oxygen (^{17}O and ^{18}O), by W. F. Giauque and H. L. Johnston at Berkeley, from Professor Joel Hildebrand in a Washington taxi. The information came in response to Urey's question, 'What is new in research at Berkeley?' Hildebrand added, 'They could not have found isotopes of a more important element', to which Urey replied 'No, not unless it was hydrogen'. That episode, recalled by Professor Brickwedde, happened two years before the discovery of deuterium.

Urey was convinced that previous spectrographic and chemical measurements of the mass of the hydrogen atom allowed for a very small amount of stable isotope of mass number 2. He based his conviction on evidence from studies of the regularities of atomic nuclei as atomic mass increased up the periodic table of elements. The 'very small' abundance relative to hydrogen of mass number 1 was calculated as around one part in 4500. This was way too small to detect from naturally or chemically produced hydrogen, and it was clear to Urey that some method of concentration was necessary in order to detect the 'heavy' isotope. He calculated that the vapour pressures of liquid hydrogen and its isotope were sufficiently different that extreme fractional distillation of liquid hydrogen would yield enough sample to analyse spectroscopically. A 2 ml sample was eventually produced in 1931 by Ferdinand Brickwedde in Washington, which yielded the required spectral line for the heavy isotope, though Urey had already detected traces of this by spectrometry in samples of commercially produced liquid hydrogen, but had delayed publication because he wanted to be sure of its presence.

The clinching experimental evidence from Brickwedde's sample was obtained late on Thanksgiving Day in November 1931:

> ...[On] that fateful day of discovery Urey arrived home hours late for his Thanksgiving dinner; the family guests had arrived, and, as his wife recalls, the only excuse he could offer was the comment 'Well, Frieda, we have got it made'.

The discovery led to the award of the Nobel Prize for chemistry in 1934 – one trusts that Frieda forgave him! A few months later, Chadwick discovered the neutron – the puzzle of exactly how elements and their isotopes were constituted was finally solved.

Stable isotope fractionation: key to palaeoclimate

From Urey and his collaborators' subsequent studies of the relative abundances of the stable isotopes of hydrogen and oxygen in water and ice and in calcium carbonate minerals, they concluded that these could not represent equilibrium at any single value of temperature. Rather the heavier isotopes were more difficult to incorporate into water vapour and in mineral precipitates at higher temperatures, a process known as fractionation, specifically as *mass-induced* isotope fractionation. This process is only noticeable and detectable in the lighter elements of the periodic table, like hydrogen and oxygen, which feature large in the climatic applications of modern stable isotope geochemistry. For example, in oxygen the mass difference between common oxygen atoms, denoted ^{16}O and its heaviest isotope, ^{18}O, is around 12%. Even so, the variation in the ratio of the abundances of the two isotopes is usually only around a few parts per thousand and it was this level of difference that Urey and his collaborators had to deal with when they adapted mass spectrometers for that purpose in the late 1940s.

In 1948 he reviewed the state of the art in oxygen isotope studies achieved at Columbia in the 1930s and since the 1940s in his new laboratories at the University of Chicago. Eighteen months previously he had come up with an idea that would turn out to be the beginnings of a new era in the science of palaeoclimatology. Here in his own words, he outlines the basics of oxygen stable isotope chemistry during the precipitation of organic calcium carbonate, together with thoughts on the potential of his new breakthrough:

> The energy [aka enthalpy] and entropy [the tendency to decrease internal order] and hence the free energy of chemical substances depend on the vibration frequency of the molecules, and these depend on the masses of the atoms [involved]... there is a slight temperature coefficient for the abundance of the O^{18} isotope in the calcium carbonate as compared with

that in the water...The temperature coefficient...makes possible a new thermometer of great durability, which may have been buried in the rocks for hundreds of millions of years after recording the temperature of some past geological epoch and then having remained unchanged to the present time.

It is evident that, if an animal deposits calcium carbonate in equilibrium with water in which it lives, and the shell sinks to the bottom of the sea and is buried securely in the earth and remains unchanged from that time to this, it is only necessary to determine the ratios of the isotopes of oxygen in the shell today in order to know the temperature at which the animal lived. This particular application of the chemical difference in the processes of isotopes occurred to me a year and a half ago, and since that time my colleagues and I have been trying to solve the severe difficult problems encountered in making such measurements of palaeotemperature.

He goes on to discuss these various difficulties: the design of a new highly accurate and precise collecting system for the mass spectrometer; the preparation of the wet chemical process to obtain carbon dioxide from calcite samples; the determination of an empirical temperature scale from modern marine shelly organisms; the careful selection of coarsely crystalline calcite samples. He then discloses the very first attempt at calculating palaeotemperatures with his new experimental set-up. The results were gained from the robust remains of Upper Cretaceous marine belemnite fossils (ancient squid-like creatures with a rigid, thick, calcite skeletal support) obtained from the Chalk of southern England as provided by the palaeontologist S.J. Stubblefield of the Geological Museum, London. The preliminary results gave mean temperatures for the shallow Cretaceous seas in the range 18–27°C. He concluded on a cautious but optimistic note:

The data thus far secured are not sufficient to draw extensive and varied conclusions in regard to past geological temperatures, but they are sufficient to lead us to believe that at least some measurements of past temperatures can be made, and hence a quantitative estimate for past climatic conditions of the earth secured...These studies can be compared to the radioactive timescale...that method is very simple in principle, but the actual carrying out of the research which established that timescale involved very careful work covering many years.

So, Urey showed great prescience and patience to take advantage of heavy oxygen in determining the composition of the natural precipitates of calcium carbonate derived from waters at various temperatures. After a decade and more of intensive study, he together with collaborators, S. Epstein, H.A. Lowenstam and C.R. McKinney, carefully laid out their stall in one of the great classics of palaeoclimate literature published in 1961 in the *Bulletin of the Geological Society of America* as the 'Measurement of paleotemperatures and temperatures of the Upper Cretaceous of England, Denmark, and the south-eastern United States'. The title to this paper is rather misleading, for the star of the show was the authors' demonstration of seasonal temperature variations in a carefully sampled Jurassic belemnite skeleton (Fig. 13.2).

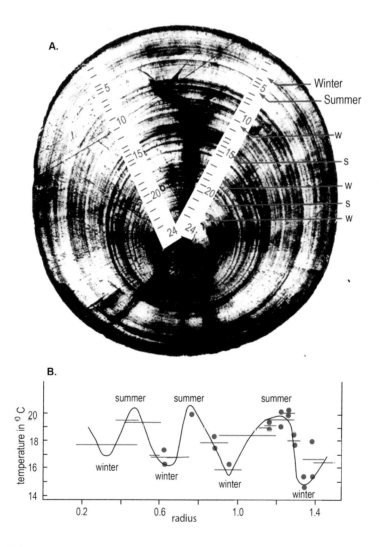

Figure 13.2 A. Urey and his team's Jurassic calcite belemnite 'skeleton' viewed in thin section showing growth rings and location of samples taken for oxygen stable isotope analysis. B. Graph to show the palaeotemperature profile derived from the mass spectrometer analyses of the samples indicated in A. The squid-like creature appears to have lived through a pleasant seasonal life while dodging plesiosaurs in the tropics for four years or so.

The carbon cycle: the Urey equation and an historical perspective

In the late 1940s and early 50s Urey had, in his own words, arrived at 'certain definite conclusions' in his research into planetology, as he called it. This was the pioneering 1952 paper 'On the early chemical history of the earth and the origin of life' published in the *Proceedings of the National Academy of Science USA*. He began:

> In the course of an extended study on the origin of the planets I have come to certain definite conclusions relative to the early chemical conditions on the earth and their bearing on the origin of life.

The conclusions involved the evidence for an anaerobic (oxygen-free), water-rich early atmosphere containing methane and carbon dioxide that existed before the advent of the global photosynthesis that enabled a steady supply and build-up of free oxygen. He asks the penetrating question:'what could reasonably be expected to convert this [primitive] atmosphere into the present one existing on the earth?'.

His answer was that on a newly crystallizing and solidifying earth's crust, atmospheric weathering of rock-forming silicate minerals would play a vital role in the evolution of a virtually carbon-free atmosphere. He envisaged the inorganic production of oxygen from the dissociation of water, which would have initially reduced carbon compounds to carbon dioxide and reduced iron to more oxidized states. He proposed that the carbon dioxide released by this process and by gaseous emanations from the mantle via volcanic activity would have been mopped up by its reaction with calcium-bearing silicate minerals abundantly present in the nascent crust – particularly the calcic plagioclase, anorthite, of the kind investigated by Norman Bowen (Chapter 8). That reaction, leading to the conversion of mineral silicate to calcium carbonate (forming limy sediments in the oceans) and silica is now referred to as the Urey Reaction. Written in its simplest form it goes:

calcium silicate (generalized) + *carbon dioxide = calcium carbonate + silicon dioxide*

$$CaSiO_3 + CO_2 = CaCO_3 + SiO_2$$

Urey wrote:

> Of course the silicates may have been a variety of [rock-forming] minerals but the pressure of CO_2 was always kept at a low level by this reaction or similar reactions just as it is now. Plutonic activities reverse the reaction from time to time, but on the average the reaction probably proceeds to the right as carbon compounds come from the earth's interior, and in fact

no evidence for the deposition [aqueous crystallization] of calcium silicate in sediments seems to exist.

Unknown to Urey (and to anyone else), he was in fact rediscovering the wheel, for 100 years previously a young French mineralogist/chemist, J.-J. Ébelmen, had pursued the balances needed to keep the atmosphere in its present state of oxygenation and carbonization – his work went unrecognized until unearthed and developed by Robert Berner in the 1970s, a topic to be expanded in Chapter 17.

The basis of life

It was also in the 1950s that Urey became associated with the man who set up one of the most famous experiments in history. This was Stanley Miller whose paper 'A production of amino acids under possible primitive earth conditions' was published in the leading American journal *Science* in May 1953. After graduating from Berkeley with a degree in chemistry, Miller had become Urey's doctoral student at Chicago in 1952 after hearing him lecture on his notion of an anaerobic pre-biotic atmosphere. This led Miller on to propose an experiment for his Ph.D. to synthesize amino acids in the laboratory. It was a timely endeavour, for the eventual publication of Miller's results took place just a month after the famous publication by F. Crick and J. Watson in *Nature* on the double helix structure of the DNA molecule. The successful run of Miller's experiment gave positive results after the experimental mix was subject to electrical discharge for a whole week. One can do no better than to quote in full the first two paragraphs of Miller's paper and to view an annotated depiction of the experimental apparatus (Figure 13.3).

> The idea that the organic compounds that serve as the basis of life were formed when the earth had an atmosphere of methane, ammonia, water and hydrogen instead of carbon dioxide, nitrogen, oxygen and water was suggested by Oparin and has been given emphasis recently by Urey and Bernal.
>
> In order to test this hypothesis, an apparatus was built to circulate CH_4, NH_3, H_2O, and H_2 past an electrical discharge. The resulting mixture has been tested for amino acids by paper chromatography. Electrical discharge was used to form free radicals instead of ultraviolet light, because quartz absorbs wavelengths short enough to cause photo-dissociation of the gases. Electrical discharge may have played a significant role in the formation of compounds in the primitive atmosphere.

It later emerged that Urey refused to have his name as co-author on Miller's publication as he had no wish to claim undue credit and, as a Nobel Prize winner, perhaps be the subject of insinuations that the results were primarily his. One sees the selfless example of Gilbert Lewis, his own Ph.D. supervisor, in such actions that were typical of the man himself.

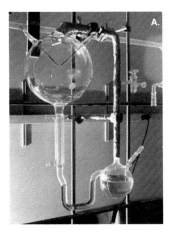

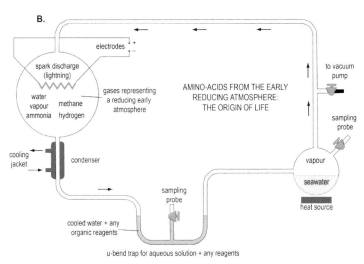

Figure 13.3 The ultimate experiment – the origin of life in a test tube. Diagram of Stanley Miller's 1953 apparatus from which he collected amino acids synthesized from a mix of gases subject to repeated electrical discharges over a week-long experimental run.

PART 6

Coda: Legacies and Connections

As I hope the reader has discovered so far, the advances made by the pioneers played a major role in creating earth sciences, with its emphasis on the processes that enable understanding of the global recycling of virtually everything: magma, plates, water, carbon dioxide, etc. It is beyond this book's scope to attempt to summarize *all* the later additions and modifications to the pioneers' work down to the present day. Instead, I have attempted in this closing part to concentrate on a few central themes advanced mostly in the later decades of the twentieth century with emphasis on the unifying threads that must run through any narrative of this kind. I do not return to details of the more technical aspects of developments in most fields. Rather I explore various ramifications of Wegener's mobilist earth that eventually led to the theory of plate tectonics, also to the relationship between tectonics, climate and global carbon cycling, finally to that greatest of confirmations – the orbital theory of palaeoclimate.

First up comes the intriguing story of how Andrija Mohorovičić's discontinuity was subsequently recognized by geologists in the fragmented and tectonized remnants of masses of former ocean crust and mantle now exposed in Alpine–Himalayan and older mountain belts – the Steinmann Trinity/ophiolite story.

In contrast to the Moho, Beno Gutenberg's discovery of the G-Zone, a change in state, not of substance, has been thrillingly confirmed in recent years by new seismological techniques with its wider importance for melt migration and for the dynamics of plate motion – the balance of applied and resisting forces that act across it.

Picking up on the wider theme of tectonics and earth cycling, we turn away from the continents themselves to the huge consequences of ocean exploration by remote geophysics (sonar, magnetism, gravity) and focus on post-World War II advances – major breakthroughs that led to confirmation of the process of sea-floor spreading and to plate tectonics.

Harold Urey's legacy to earth cycling via his weathering equation has been reinforced and joined with the plate-tectonic driver to explain the global carbon budget and, as a direct consequence, that of large-scale climate changes, in particular the hitherto unexplained trend towards oceanic cooling over the past 50 or so million years.

Urey's pioneering work on oxygen stable isotopes and palaeotemperatures has been followed by precise evidence for glacial/interglacial conditions during the last

4 million years gained from spectrometric measurements on calcareous microfossils found in ocean floor sediment cores. This has enabled estimates for global sea-level changes and to statistical confirmation of Milanković–Croll orbital theory over deeper geological time.

14

Mohos exposed

Over the five or so decades since the crust–mantle boundary was remotely detected by Mohorovičić, countless earthquake seismograms, augmented by information from nuclear testing, established as a fact that the Moho is a universal feature of the planet's make-up. It underlies both oceans and continents alike, despite huge difference in crustal thickness and composition between the two realms. There is thin crust under the ocean floor, just a few kilometres, but deeper under the continents, usually 30–35 km – greatest of all, up to 60 km, under the Tibetan Plateau.

Steinmann's association

Indirect knowledge concerning the nature of the ocean crust, Moho and mantle came not from the interpretation of geophysical soundings but from field geology. It was in the Alpine–Himalayan mountain chain that geologists noticed and described the not-infrequent occurrence of bodies of attractive, softish, green-banded rock known as serpentinite. Its chemical composition, rich in magnesium, left no doubt as to its origin by hydration reactions with forsterite olivine, a major component of peridotite, whose greater density and seismic wave travel time compared to granitic and basaltic rocks closely identified it as the chief mantle constituent. By the 1930s it had become obvious to Norman Bowen and others that such a dense, refractory ('barren') rock could never make more magma or lava flows but must have crystallized out as a residue from partial melting.

What was even more intriguing was the association of serpentinite with two other distinctive rocks. One was basalt moulded into pillow-like masses whose origins had engendered considerable controversy before field observations of active volcanic eruptions early in the twentieth century. These showed that they formed by basaltic lava extruding undersea – the lava masses chilling and sagging over each other as the flows advanced like dropped bags of wet cement. The other rock was of sedimentary origin – successions of often reddish siliceous material known as chert occurring in successive strata, each usually just a few decimetres thick. In thin section on the microscope stage these bore traces of their formation from accumulations of siliceous planktonic creatures – radiolarians – common enough as siliceous oozes in the deep and fertile parts of modern oceans and with a long geological history stretching back deep into the Palaeozoic era over 400 million years ago.

One Austrian geologist in particular, Gustav Steinmann, published extensively on this mysterious and colourful assemblage of contrasting rocks from his field mapping in the Austro-Italian Alps and Apennines from the early 1900s into the 1920s. Here are the words of modern Swiss and French workers in their tribute to him (Bernoulli et al., 2003):

> In 1905, Gustav Steinmann noted the close association of serpentinites, diabase [basalt] and radiolarite and considered this 'greenstone 'or ophiolitic association [ophiolite: a snakeskin-like rock: Greek: *ophis* - snake] as characteristic for the axial part of the 'geosyncline' [see later] and the deep ocean floor...

Enter Edward Bailey

It was that doyen of the British Geological Survey, Edward Bailey, who in the 1930s first realized the global occurrence and tectonic significance of this weird assemblage of rock types recognized by Steinmann. Bailey, a heroic veteran of World War I and an assiduous and eccentric field worker in the ancient Caledonian mountain belts of Scotland (see Fig. 9.5), also undertook field visits to Europe and had read widely in the German, French and Italian literature of Alpine–Himalayan geology. In 1936 the Geological Society of America published a paper read by him in September of that year at the *Tercentenary Conference of the Arts and Sciences* at Harvard University. He was then in his last year at the helm of the Department of Geology at the University of Glasgow, the next year becoming Director of the British Geological Survey. Bailey's wide-ranging review was entitled 'Sedimentation in relation to tectonics', and from the perspective of today might be said to encompass all the various strands of this great geologist's fertile, creative and playful mind. In it he introduced the concept of the '...architecture of the Earth's crust...' as providing '...the subject matter of Tectonics.' and enlists the help of the young discipline of sedimentology and the much older one of palaeontology in achieving his aims in this interdisciplinary field. After a discursive section on recent advances in the use of sedimentary structures in determining the 'way-up' of tectonized strata (in overfolds known as nappes) he revealed his own stance in fluent prose – he was an obligate Wegenerian:

> I have already pointed out that the deformation of folded mountains indicates some form of lateral compression. There is no general agreement as

to cause. The oldest theory, that of Elie de Beaumont [1798–1874], attributes folded mountains to pressure in the earth's crust through accommodation due to a cooling, shrinking interior. Despite radioactivity, this view still has important advocates. Other theories invoke some type of glacier-like motion in the crust. Others picture the process rather in metaphors of ice packs and ice bergs. Here we find Wegener's theory of continental drift, which, in spite of its difficulties, furnishes, to my mind, the only rational explanation yet offered of the apparent dispersal of the Late Carboniferous, or Early Permian, glaciation of Gondwanaland. How else can we account for the records of this glaciation in India, South Africa, Australia, Brazil, and the Falkland Islands?

Indeed!

He then gave an overview of deep-water sedimentary deposits, chiefly involving graded bedding (an upward decrease in grain size, denoting deceleration during deposition) and what we now call debris flows and turbidites – deposits of turbidity currents. He wrote an engaging section on deep-water submarine slopes and carried on introducing Steinmann's legacy of discovery:

> I hope that I have persuaded some of my audience to follow me down the geosynclinal slopes, in company with earthquake-shaken grit and sand, to greater depths than are reached by ordinary bottom currents. Others have ventured in advance. For instance, W.H. Twenhofel, in the 1926 edition of his *Treatise on Sedimentation*, has written: 'About many ocean islands and the younger continental margins are steep slopes down which materials are readily moved to the deeper waters of the bathyl environment. Such steep slopes lead to inclined deposition and favour slumping, particularly if the steep slopes are the loci of seismic movement'. Now, with some hesitation, I suggest we take the final plunge, and greatly daring, descend into the abysses explored by G. Steinmann.

> In 1889, Steinmann visited the Iberg klippes east of Lake Lucerne, and found two closely associated suites of rock, hitherto unrecognized in northern Switzerland: the one consists of radiolarian chert and brown-red clay, the other of ophiolitic igneous material…The sediments are of Upper Jurassic and Lower Cretaceous date; but they contrast strongly in character with neighbouring exposures of the same age. In fact, they afford a striking example of exotic facies…

> Before Steinmann had reached this stage in his tectonic education, he had already played a leading role in establishing the association of radiolarian chert and ophiolites in many other parts of the world, irrespective of date.

Again and again, these rocks are together in the middle belts of geosynclines, ancient and modern.

A few words are in order here on the term 'geosyncline'. It was the invention
of nineteenth-century North American geologists (notably James Dana in 1873)
working in the Appalachians. They envisaged the predecessor to that long, sinuous
and beautiful mountain range as a former trough-like tract, a sedimentary basin,
whose active and rapid subsidence enabled the entrapment, often in deep water, of
thousands of metres' thickness of detritus over hundreds of millions of years. As such
it contrasted greatly with the much slower subsidence and accumulations of sediment
found on equivalent strata (as determined by fossil content) on continental 'platforms'
in adjacent Mid-Western states like Iowa. Subsequent adaption of the geosynclinal
concept to mountain belts across the world led to a lexiconic orgy of name-calling –
miogeosynclines, eugeosynclines, zygogeosynclines, etc., all wearisomely familiar to
me as an undergraduate student. Their merciful extinction in the late 1960s and early
1970s came in the face of direct and real analogies from modern continental margins
provided by plate tectonics, notably synthesized in classic papers by Andrew Mitchell
and Harold Reading.

Towards the end of his Harvard address, Bailey impresses one with his foresight
when he linked the work of Steinmann, the great Swiss Alpine geologist Émil Argand,
and the Wegenerian concepts of continental drift:

> E. Argand has developed Steinmann's views, and has pictured geosynclines
> as determined by stretching, by continental drift-apart, which attenuates
> the sial layer [continental crust] and eventually allows sima [new ocean
> crust] to reach the bottom of the sea at bathyl or abyssal depths. If such
> drift-separation continues, a new ocean bed may be developed, covered
> with products of submarine eruptions. If, as has thrice happened in the
> post-Cambrian history of Europe, a drift of separation gives place to a drift
> of approach, then a folded mountain chain comes into being.

Bailey had seized on Argand's vision, bringing us forward to modern (post-1969)
concepts of tectonic recycling with the now-vanished Iapetus, Rheic and Tethyan
oceans giving us the Appalachian–Caledonian, Variscan and Alpine–Himalayan
mountain belts respectively. At the very end of his address, he left his audience with
a ringing modern analogue from the Pacific Ocean to illustrate his thoughts on how
the crustal compression of geosynclines during the elimination of these ancient oceans
might include fragments of ocean crust, volcanic edifices and sediment:

> Imagine a thrust-mass advancing on the floor of the Pacific and meeting
> a Hawaiian island. It might push it forward. It might climb its slopes. At
> any rate, it would have to take notice of the obstruction. I am doubtful

whether this aspect of the problem has been sufficiently considered in recent interpretations of the Pennine Alps.

Much later, in retirement in 1953, Bailey, together with William McCallien, former Professor of Geology at the University of Ankara, drew attention to the occurrence in central Turkey, of just such an assemblage – broken-up masses known as 'mélanges' (French: *mixtures*) with common pillow lavas:

> In many cases these lavas occur with serpentinite and radiolarian cherts in what is known as the green-rock or ophiolite association. Thus the recognition of pillow lavas completes for Turkey the Steinmann trinity… [his] realization of this trinity, as a world-wide, age-long characteristic of geosynclinal deposition ranks, among the most exciting achievements of geological research…

So was introduced the term 'Steinmann Trinity' into mainstream geology. The relevant fact acknowledged by Bailey was that it featured evidence for former deep-water oceanic magmatism and related sedimentation, the assemblage eventually brought up during mountain building by thrust faulting, the former ultrabasic peridotite changed in rock volume and density to serpentinite by hydration during episodes of prolonged stratal shortening. Bailey had recognized such rocks in his beloved Caledonides – at Ballantrae, Ayrshire and along the Highland Border to the Grampians. Largest and most spectacular was the example of the Lizard Peninsula in Cornwall, located in the Variscides of SW England where Henry Dewey (uncle of John) and John Flett (see Chapter 9) had, decades earlier, recognized their general significance for the magmatic origin and aqueous alteration of basaltic pillow lavas.

Olympian secrets revealed

It was the island of Cyprus in the eastern Mediterranean that provided the full story behind the usually tectonized and fragmented examples of the Steinmann Trinity. In 1963, Ian Gass of the University of Leeds and D. Masson-Smith of the British Geological Survey published their now classic account of the geology and geophysics of the Troodos Massif (Fig. 14.1), a mountainous range of silica-poor (basic and ultrabasic) igneous rocks that had formed both as magmatic intrusions and volcanic extrusions. The core of the massif, Mount Olympus (Fig. 14.2), seat of the goddess Aphrodite according to Strabo, was a plutonic complex and the site of a large positive gravity anomaly. The latter was due to underlying high-density ultrabasic rocks, ranging in surface outcrops from olivine-rich examples (dunites and peridotites) in the central part to a surround of basic and intermediate types (gabbros, granophyres). These field relations and the evidence of the anomaly pointed to the whole massif as being underlain by a 'vast mass of underlying, high-density ultrabasic material'

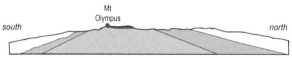

Figure 14.1 The mantle risen up – a section through the oceanic mantle and a version of the igneous geology of the Troodos massif, Cyprus, with a north–south vertical section through Mt Olympus to show its broadly domal structure (after Gass and Masson-Smith, 1963).

Figure 14.2 A wintertime, snowy Mt Olympus, the plutonic lower oceanic crustal core of the Troodos Massif, Cyprus. Wikimedia Commons.

Figure 14.3 A. Close-up of part of a single pillow of olivine-rich basalt lava that once formed the sea floor to the mid-ocean ridge that generated the Troodos ophiolite from outcrops near Akaki Bridge, Klirou. Plentiful millimetric brown-weathering olivine crystals may be seen on fresh rock surfaces. The hole next to the lens cap is where a rock drill has sampled a cylindrical core of the basalt for its palaeomagnetic signature. Pillow lavas from Troodos are also intruded by sheeted dykes, as seen in Figure 9.7B. B. Sheeted dolerite dykes (dykes intruded one into another) seen here in a road cut near Palaichora in the Troodos massif are part of the former middle part of Tethys Ocean crust. Each younger dyke is 'chilled' against its neighbour, having quickly solidified in contact with its older, cooler and crystalline neighbour, its silicate minerals having markedly reduced crystal size and the rock a darker colour.

estimated to be around 12 km thick and whose progressive crystallization resulted in the observed gradual upward and outward zoning into basic and intermediate intrusions.

Rimming the plutonic core there was first a spectacular N–S orientated dyke swarm (Fig. 14.3B), individual dykes a few metres or less in thickness and often 'sheeted', the dykes intruding each other by turns, the whole formation 2–4 km thick. Above them and often crosscut by occasional dykes (see Figure 9.7B) were a *c.*1 km thick sequence of basaltic pillow lavas (Fig. 14.3A) and associated radiolarian cherts.

During their exhaustive and sometimes dangerous fieldwork (it was a time of insurrection by Greek Cypriots against the occupying Brits) over several years, Gass and Masson-Smith had unearthed the full, three-layered structure of the oceanic crust and its underlying mantle. It was there because thrust tectonics had brought it up to the surface during the closure of portions of the ancient Tethyan ocean (Fig. 14.4). In their own words:

> It is proposed that the Troodos massif evolved in pre-Triassic times as a volcanic pile in an oceanic environment….it is tentatively suggested that the exposed rocks of the Troodos massif might indicate the structure and nature of other oceanic volcanic masses and that the exposed rocks of the Troodos Plutonic Complex might in fact represent the sub-Mohorovičić peridotite material which has been partially fused to provide the volcanic

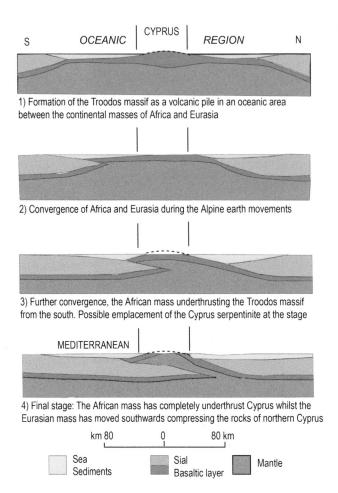

1) Formation of the Troodos massif as a volcanic pile in an oceanic area between the continental masses of Africa and Eurasia

2) Convergence of Africa and Eurasia during the Alpine earth movements

3) Further convergence, the African mass underthrusting the Troodos massif from the south. Possible emplacement of the Cyprus serpentinite at the stage

4) Final stage: The African mass has completely underthrust Cyprus whilst the Eurasian mass has moved southwards compressing the rocks of northern Cyprus

Figure 14.4 Version of Gass and Masson-Smith's 1963 summary of their conclusions concerning the origin and development of the Troodos Massif – the first attempt at relating ophiolites to their ocean crustal origins.

material of the Sheeted Intrusive Complex and the Pillow Lava Series. It is also thought that whilst this sub-Mohorovičić material was either partially or completely fused it was mobile enough to allow differentiation into a roughly stratiform complex ranging in composition from dunites and peridotites to overlying gabbro and granophyre.

In 1968, at the height of the 'plate revolution', Ian Gass, still in Leeds at the time though destined soon to occupy (with distinction) the founding Professorship of Earth Sciences at the Open University, published a *Nature* paper asking a question as his title, 'Is the Troodos Massif of Cyprus a fragment of Mesozoic ocean floor?' He replied in the affirmative, reiterating and updating his 1963 effort with Masson-Smith, but still with a good deal of detailed indecision concerning the exact nature of that ocean floor. Was it a portion of a mid-ocean ridge, a flanking volcanic centre or, as

Edward Bailey had fancied, an errant sea mount or guyot caught up in the maelstrom of continental collision or subduction? Such questions are still asked by geologists working today in the Troodos. At any rate, Gass concluded that the plutonic rocks were certainly Tethyan upper mantle, the Moho boundary was at its junction with gabbro and sheeted dykes as Layer 3 of the ocean crust, the pillow lavas erupting to form the ocean floor.

The process of thrusting that caused the Troodos massif to be formed and preserved during closure of the Tethys is now known as part of the general process of 'obduction'. The fragmentary and incomplete record of oceanic structure contained in most examples of such ophiolites is attributed to their systematic dismembering during the thrusting process that emplaced them during ocean closure. The Troodos example, whose investigation was followed by that of the equally spectacular and widespread Semail ophiolite in Oman by Mike Searle and others, are an heroic exception.

15

Slip-sliding away

The seismic evidence advanced by Beno Gutenberg for the existence of the G-Zone between 1926 and 1959 was largely confirmed by further studies (including contributions by Inge Lehmann) in the late 1950s and early 60s. Answering the question of what might cause such a feature has occupied the minds of numerous subsequent researchers. That the topic is a fundamental one was clear at once to the early proponents of plate tectonics – steady motion of the lithosphere upon flowable asthenosphere must involve a balance of an asthenosphere resisting force against an applied lithospheric driving force.

G-Zone 'bounce points'

Numerous explanations for the G-Zone have been forthcoming: that it might be due to a simple increase of temperature; that the presence of water might cause mineral hydration (olivine to serpentinite); that it might represent a change in composition, crystal size, crystal orientation (anisotropy) and/or the degree of crystallinity. As we have seen, Gutenberg himself eventually favoured the explanation proposed first by Wolff, that it might represent a zone of partial melting due to the increase of temperature with depth just intersecting the mantle rock melting curve (as in Fig. 5.5).

Researchers over the past two decades, notably Peter Shearer and, more recently, Catherine Rychert, have returned to the topic of the G-Zone with great gusto. In 1991, Shearer had demonstrated deeper mantle discontinuities at 410, 520 and 660 km depth, widely believed to be due to changes in olivine crystal structure to denser forms brought about by extreme pressure at such depths. He was able to demonstrate these by making use of long-period S-wave arrivals from surface-reflected waves detected in the massive databases of the Global Digital Seismic Network. These were revealed by new techniques of data analysis by 'stacking', to average out local statistical 'noise' of various kinds in the seismic signals.

Applying such techniques to search for and interpret the shallower G-Zone, Rychert and Shearer in 2009 (with a concise coda by N. Schmerr in 2012) studied the records of S-waves traversing the Pacific crust and mantle. A significant discovery was the generation during shallow (<60 km) earthquakes of S-waves whose curved passage through the upper mantle to the crustal surface and then, by reflection, to equidistant seismology receiver stations led to an interesting scenario (Fig. 15.1). This was the generation at surface reflection points ('bounce points') of both high-amplitude surface reflection arrivals, coded as SS ray paths, and by preliminary arrivals attributed to deeper underside reflections from the relatively thin (*c.*15 km or so) and abrupt G-Zone, coded as SdS rays. Having a shorter travel distance (at least a minimum of two lithosphere thicknesses), the low-amplitude SdS bottom reflections arrived at receiving seismometers well before those of the main SS reflections – (around 25 seconds) – hence their status as 'precursor arrivals'.

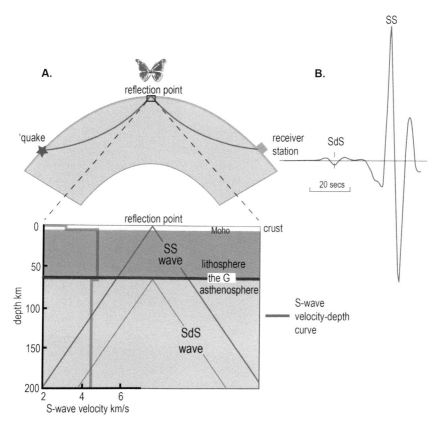

Figure 15.1 A. Versions of diagrams after Schmerr (2012) to illustrate the principles behind the use of shear wave reflections to identify and map the G-Zone beneath oceanic lithosphere. B. The nature of the seismic signal recorded at a receiver station showing the precursor arrival of a G-reflected SdS wave.

An active G-Zone?

Results from the Pacific enabled determination of the thickness of the lithosphere over the whole ocean, revealing its thickening with age, confirming the indirect results from heat flow measurements and models of conductive cooling made 30 years previously. According to Rychert and her colleagues in 2020, the physical nature of the G-Zone most probably involves a state-change such as a degree of partial melting rather than just a temperature change or crystallographic effect. In this way the conclusions of Wolff and Gutenberg 100 years ago seem amazingly prescient. Regarding the wider significance of the G-Zone, its detection deep (up to 200 km) under the ancient Archaean continental shields suggested that it is not just the passive layer implied by the term 'discontinuity' but constitutes a more active part of the mechanism that allows deep planetary heat to be released from convecting asthenosphere into the lithosphere (Fig. 15.2). It accompanies both melting during rise of asthenosphere under mid-ocean ridges (see below) and thickening of the lithosphere by underthrusting, as below the Tibetan Plateau.

On a smaller scale, the occurrence of numerous isolated oceanic volcanic clusters (e.g., the Galapagos Islands), volcanic chains (Hawaii), mid-ocean ridge islands (Iceland) and developing continental rift systems (Afar and the Red Sea) implies the rise of mobile melt from the G-Zone. These features occur under plate boundaries and under developing continental rifts. They may be due in some way to upwellings from small-scale convection cells in the asthenosphere.

Partial melting under mid-ocean ridges and developing continental rifts occurs when rising asthenosphere partially melts because of an increase in temperature caused by decompression as geostatic pressure decreases. This thermodynamic heat, well-known to Norman Bowen, cannot be entirely lost by conduction into adjacent static mantle, since this is a slow process compared to the rate of rise engendered by the

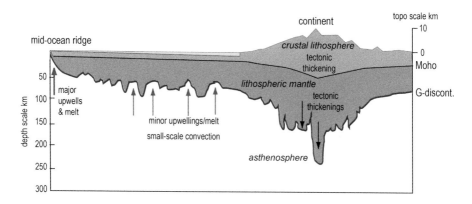

Figure 15.2 Version of the Rychert et al. (2020) synthesis of the extent and significance of the G-zone.

driving forces of plate motion. Such heating up of rising material is known as adiabatic ascent, also a phenomenon in rising air masses and well-known to meteorologists like Alfred Wegener. Melting and magmatism is also produced by the frictional heat and lowered melting temperatures induced by the descent of hydrated oceanic mantle lithosphere (it is serpentinized) back into the asthenosphere at and above subduction zones. It is responsible for the 'ring of fire' volcanic arcs of the circum-Pacific margins.

Doing it for themselves

From the time of Arthur Holmes down to the late 1960s the driving mechanism for drift was attributed to the drag exerted by mantle convection currents. Despite the persistent opposition of Harold Jeffreys to any deviation from his stiff and unyielding mantle, rather like himself by all accounts, the equally persistent and stubborn Gutenberg insisted that a low seismic velocity zone bounded a lid of lithosphere over convecting asthenosphere. His views won the day when in the early 1960s Keith Runcorn used the Rayleigh Number, the simple ratio of heat transfer by conduction to that by convection (with its dominance by a viscous term), to give a clear message that the asthenosphere could in fact slowly convect – it could drive the continental drift by viscous drag on the base of the lithosphere.

But a problem soon became apparent – long-known experiments on the convection of viscous fluids by Henri Bérnard in 1900 and later by Rayleigh had shown that the length scale of convection cells formed by uniform heating from below were too small in relation to the lateral extent of large lithospheric plates. So, an often-asked question in geophysics tutorials at Leeds in the early 1970s was: 'What drives continental drift and plate motions if not convection?'

Donald Forsyth and Seiya Uyeda set out to answer this question in their 1975 paper, 'On the relative importance of the driving forces of plate motion' published in the *Geophysical Journal of the Royal Astronomical Society*. They pointed out that although the theory of plate tectonics had been remarkably successful in explaining the major features of the mobile outer earth, the theory was incomplete, concentrating on the relative motion of plates rather than on possible driving mechanisms for that motion. In the language of physics, it was mostly a kinematic theory (Greek: *kinema*; motion) that awaited a full physical explanation and dynamic input in terms of the forces involved. They pointed out that a successful theory had to be capable of explaining: 1) the magnitude of the mechanical energy needed to move the plates as measured by the annual release of elastic energy by earthquakes; 2) the stress distributions measured within plates and at their boundaries; 3) the plate velocity vectors established by marine magnetic anomalies.

Forsyth and Uyeda proceeded by what is known as an inverse solution, working backwards from observational data on plate motions in order to determine the relative magnitude of the forces involved. Various forces were identified that act on plates in contact with the asthenosphere and along the junctions of adjacent plates. At the

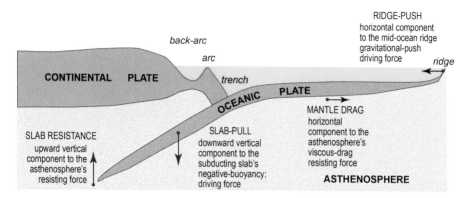

Figure 15.3 Version of Forsyth and Uyeda's 1975 diagram to illustrate the applied slab-pull and ridge-push forces and the asthenosphere's resisting forces that define the dynamics of steady plate motion.

G-Zone itself, a sliding, shear-couple arrangement is maintained (Fig. 15.3). Using the mean velocity of each of the world's plates they found that the presence of a downward-moving (subducting) slab, generating what they termed a 'slab-pull' force, was the strongest control on rates of plate motion – witness the much greater velocity of the Pacific plate compared to that of the Atlantic plate where 'ridge-push' was the only driver. It seems that whilst the asthenosphere does indeed convect to lose earth's internal radiogenic heat as Holmes, Gutenberg, Hess and Runcorn proposed, it does so on a relatively small scale – but lithospheric plates drive themselves!

16

Unzipping the oceans

The mariner's tale

It was largely the remote geophysical exploration of the ocean floor and its underlying crust and mantle that led to the spectacular advances seen in global tectonics during the 1960s and 70s. Chief among the early geomariners was Harry Hess, initially drawn to the study of geology whilst a beginning electrical engineering undergraduate at Yale. Switching courses, he graduated without distinction in 1927 and, after a couple of years working as an exploration geologist in Southern Rhodesia, took up a Ph.D. at Princeton on the field relations and petrology of an ultrabasic peridotite intrusion in the state of Virginia. Graduating in 1932, his fascination with such rocks led on to field and laboratory studies of the Stillwater layered intrusion in Montana. By the late 1930s he had become an expert on the geochemistry of the pyroxene group of silicate minerals which, along with olivine, became an integral part of the peridotite story that subsequently unfolded. His National Academy of Sciences obituarist, H.L. James, summarized his subsequent lifelong quest thus:

> His was a rare, perhaps unique talent. It combined far-ranging interests and a brilliant intuition with a capability and willingness to carry out work calling for extreme detail and precision. His career was an extraordinary one: a mineralogist of world repute who became even better known for introduction of new concepts on the origin of continents and oceans…a quiet and unassuming scientist of puckish disposition who became a wartime Navy commander and rose ultimately to the rank of Rear Admiral…

> Peridotite – its origin and significance – probably was the object of Harry's deepest scientific devotion…he realized that this mantle-derived rock was

a key to the understanding of the deeper crustal structure of the earth and to the recognition of older orogenic belts...the role of this unusual rock was to be a critical factor in his later theories of the nature and behaviour of the ocean crust.

Let us follow this later and wider career in some more detail. By the early 1950s Hess had a unique take on the problems associated with the origins of the ocean basins. During his 48 years, he had traversed the Caribbean by submarine when a young graduate student, measuring variations in gravity under the direction of Felix Vening Meinesz. During the war years of 1942–45, he had searched, using sonar, for Nazi submarines in the Atlantic and then traversed the Pacific, commanding an attack landing-support ship in the Pacific 'island-hopping' campaign against the Japanese. In both oceans his extensive sonar records had given him a unique insight into oceanic bathymetry, including the data that led to his first account at war's end of the remarkable Pacific flat-topped mountains that he named 'guyots' after the nineteenth-century geographer A.H. Guyot, proposing they were subsided former volcanic islands. By this time, he was perhaps the first land-trained geologist and mineralogist to turn to the oceans in order to reveal the secrets of the continents. His exegesis of these would dominate efforts in understanding the earth's outer structure and workings over the next fifteen years.

The first inkling of the way his thoughts were developing at Princeton came with a short contribution to a Royal Society of London discussion symposium on oceanic matters published in 1954. He began his contribution with a well-aimed message, partly directed at himself, concerning the temptation for marine geophysicists to just collect instrumental data for its own sake:

> Compared to the pre-war era there has been a great increase in the amount of geophysical work at sea, and a correspondingly vast amount of new information acquired. The energy and ingenuity of the scientists concerned has gone largely into development of new techniques and the acquisition of data, with comparatively little time spent in meditation on the broader aspects of the meaning of the results. The ideas presented here are the products of a few weeks thoughtful consideration, whereas a year would have been a more appropriate interval to do the data justice. I hope this will be borne in mind by the reader.

He highlighted 'the most momentous discovery since the war' as the determination from seismic records of the depth to the Moho, the fact that it rises from around 35 km under the continents to only around 5 km below the ocean bed. He linked this with the petrological discovery from basaltic volcanoes of masses of the dense crystalline ultrabasic rock, peridotite, brought up from depth in their source regions of the upper mantle (mantle nodules). Their presence supported evidence gained from seismic velocity calculations that peridotite must form the upper mantle beneath both the oceans and the continents. Using average rock density and thickness estimates, he calculated that an isostatic stress equilibrium should exist under both oceans and continents at depths of around 40 km.

He then addressed the question of the origin of oceanic ridges, defining three types with brief hypotheses for their origins. They were volcanic ridges like the Hawaiian chain, the early development phase of certain island arcs and mid-ocean ridges like those in the Atlantic. For the latter (Fig. 16.1) he presented a more detailed hypothesis, acknowledging the legacy of Arthur Holmes, of their origin by convectional upwelling and melting:

> It involves brecciation of the peridotite substratum by great masses of basaltic magma perhaps over an upward convection current in the mantle. Some blocks of peridotite engulfed in basalt may be present at the surface as perhaps is the case in St Paul's Rock. The somewhat lower density of this column as compared to the columns either side of it permits its surface to rise well above the ocean floor. During the time when some molten basalt was present and the temperature of the column as a whole was higher

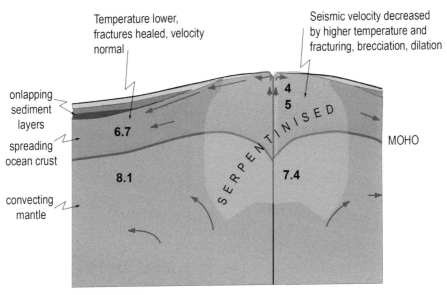

Figure 16.1 Version of Hess's 1962 view of convective motions causing the lateral movement (spread) of ocean crust and mantle about the central axis of a mid-ocean ridge. In his own words: 'Diagram to represent (1) apparent progressive overlap of ocean sediments on a mid-ocean ridge which would be the effect of the mantle moving laterally away from ridge crest, and (2) the postulated fracturing where convective flow changes direction from vertical to horizontal. Fracturing and higher temperature could account for the lower seismic velocities on ridge crests, and cooling and healing of the fractures with time, the return to normal velocities on the flanks.'

and hence less dense, the ridge might have stood high above sea level. An upward convection current beneath it would also tend to lift the ridge and cessation of the current to let it subside.

A subsequent section drew attention to the phenomenon of the hydrous alteration of olivine to serpentine in peridotite rocks. He pointed out the exothermic (heat-producing) nature of the reaction, the water involved coming out of the mantle, as a possible contributor to ridge heat flow as well as the accompanying 20% reduction of density helping to cause ridge elevation over that of the markedly deeper adjacent ocean basins.

Such were Hess's preliminary conclusions in the early 1950s as to the structure and formation of the ocean crust and their mid-ocean ridges. His further meditations over the next few years were brought together into the scientific literature from a longer document prepared for the Office of Naval Research in 1960. His colleague there, Robert Dietz, was also exploring the ideas revealed by Hess concerning oceanic structure and process at this time and published a paper on the subject in *Nature* in 1961.

Hess's own paper, *History of Ocean Basins*, was published in November 1962 as part of a volume published in celebration of the work of retiring Princeton petrologist, Arthur 'Bud' Buddington, a scientist and teacher whom Hess had greatly admired since his student days at Princeton. He began his paper in his usual frank, humorous and self-deprecating way:

> The birth of the oceans is a matter of conjecture, the subsequent history is obscure, and the present structure is just beginning to be understood. Fascinating speculation on these subjects has been plentiful, but not much of it predating the last decade holds water…I shall consider this paper as an essay in geopoetry. In order not to travel any further into the realm of fantasy than is absolutely necessary I shall hold as closely as possible to a uniformitarian approach…

His chief conclusions reached would lay the basis for what was to emerge over the next couple of years – the concept of sea-floor spreading and its rapid adoption as the basis for the description of plate tectonics, whose emergence Hess would proudly witness before his death in 1969. He envisaged a slowly convecting mantle, the currents rising under the mid-ocean ridges bringing up hot and fractured mantle material to the surface, thereby causing observed high rates of heat flux, topographic elevation above the surrounding abyssal plains and slowed-down seismic wave velocities. The rising mantle material, suitably hydrated by mantle-derived water, reached a temperature of 500°C at a depth of 5 km and formed most of the ocean crust – the serpentinized peridotite cooling, contracting and moving outwards on rift flanks as a uniform layer around 5 km thick at an average velocity of around 10 mm per year (from the evidence of known sedimentation rates and stratigraphic ages of sea floor sediment). These facts indicated that the mid-ocean ridges were ephemeral features existing for 200–300 million years, defining the life of a single active convecting cell. The only trace of an inactive former cell (taken to be Mesozoic in age) was present along a Pacific rise found by Hess in wartime and defined by plentiful extinct volcanic seamounts, atolls and guyots stretching in an arc from the Marianas trench in the west towards the Chile trench in the east.

Following Holmes, Hess deduced that continental drift occurred when rising convective mantle material came up under existing continents – these are fragmented and the constituent parts move away from each other at a uniform and identical rate on either side, so as to create a median ridge, like the Atlantic ridge system separating the Americas from W. Europe and Africa. Regarding their motion, he envisaged that the moving continental masses '…are carried passively on the mantle with convection and do not plough through ocean crust.'

At the downward moving parts of convecting mantle, he proposed that the leading edges of the continents are strongly deformed, the ocean crust buckling down, heating up and losing its serpentinite water to the ocean. Its cover of sediment and decoration of volcanic seamounts and guyots 'ride down into the jaw crusher of the descending limb, are metamorphosed, and eventually probably are welded onto continents.'

He concluded that the ocean crust and the ocean basins themselves are an impermanent surface feature of a dynamic planet, but that at the same time the continents are permanent – the destruction of the one causing the 'welding-on' growth of the other. Thus emerges a cyclical scheme of oceanic birth, life and death at the same time as continental growth – a worthy successor to James Hutton's original scheme. In Hess's modest words:

> The writer has attempted to invent an evolution for ocean basins. It is hardly likely that all of the numerous assumptions made are correct. Nevertheless, it appears to be a useful framework for testing various and sundry groups of hypotheses relating to the oceans. It is hoped that the framework with necessary patching and repair may eventually form the basis for a new and sounder structure.

And it did just that. The 'patching and repair', as Hess put it, rapidly progressed over the next seven years or so. Something of the nature of Hess as a person is revealed at the end of his obituary:

> Harry's trademark, always evident in his doodles, was the rabbit. But, as many found, this slightly built, quiet, unobtrusive man was no rabbit in fact; he was a fierce fighter for science, a dedicated and steadfast defender of any cause he thought to be just. Yet withal he was a gentle and kindly person, tolerant of the foibles and weaknesses of mankind – including his

own. Those of us who knew him lost a great friend, and the world lost a great scientist and a scientist-statesman.

As we re-read Hess's work today with the benefits of hindsight, we can identify several avenues of research that were stimulated by it. At the beginning of the 1960s, many of his hitherto unpublished ideas concerning oceanic tectonics had already been widely available, as another paragraph from his obituary reveals:

> Harry Hess was a pioneer in development of the now widely accepted theory of ocean-floor spreading. In 1960, in a widely circulated report to the Office of Naval Research, Harry proposed that the mid-ocean ridges were the loci of upwelling mantle convection cells that progressively moved mantle material outward and eventually under the continents, a brilliant concept that now appears to be confirmed by the symmetrical distribution of magnetic anomalies on both sides of the ridges. His paper was published formally in 1962, and a study made in 1969 indicates that it was the most referenced work in solid-earth geophysics in the years 1966–1968. Whether his concept of a serpentinized peridotite under the ocean floors proves valid or not, this paper stimulated intense research and is part of what is the major advance in geologic science of this century.

The appearance in 1963 of the first indirect, but conclusive, remote-sensed evidence for sea-floor spreading was provided by Fred Vine and Drummond Matthews – it must have delighted Hess. Their discovery and interpretation of symmetrical normal and reversed magnetic anomalies measured over the Atlantic mid-ocean ridge was ascribed to Hess-style symmetrical sea-floor spreading. Subsequent detailed sea-floor mapping and dating of the reversed and normal magnetization in the volcanic pillow lavas of Layer 2 of the ocean crust enabled a detailed record to be established of the history of spreading vectors for all of the world's ocean crust.

As a matter of fact, symmetrical negative and positive magnetic anomalies and positive gravity anomalies about the median oceanic depths of the Red Sea and Gulf of Aden had already been recognized by Ron Girdler in 1958 and 1961 but he did not develop the implications of their detailed symmetry about the central rift. Nevertheless, he interpreted them as due to convective cells of basaltic igneous intrusion by dyke-like forms. He wrote:

> It is seen that the trend of the anomalies is almost exactly parallel to the sinuous shorelines of the Red Sea, suggesting that the continental crust has been torn apart leaving room for the intrusion of basic igneous rocks which give rise the large magnetic anomalies.

And that:

the axial, deep water is probably underlain by a series of wide, parallel fissures filled with basic igneous rock. [i.e., dykes]

This is pure Holmes of 1944 vintage, a book which Ronnie must have read whilst an undergraduate at the University of Reading. It was this process of basic magmatic intrusion above convective cells that also led him to conclude in his 1962 *Nature* paper, 'Initiation of continental drift', that the young (*c*.30 million-year-old) continental separation of Arabia from Africa and of similar processes in the Gulf of California could be entirely explained by crustal extension, 'the formation of new oceanic crust depending on the degree of extension', as he puts it in the concluding sentence to his paper.

A key development by Hess himself was the link between convecting mantle, ridge crustal production and the lateral movement of this crust away from the ridge:

The Mid-Atlantic Ridge is median because the continental areas on each side of it have moved away from it at the same rate – 1 cm/yr. This is not exactly the same as continental drift. The continents do not plow through oceanic crust impelled by unknown forces; rather they ride passively on mantle material as it comes to the surface at the crest of the ridge and then moves laterally away from it.

It is these last six words that signify most, for it was unclear how Hess's new 5 km-thick ocean crust would have carried on moving once the lateral limits of the uplifting ridge mantle were exceeded.

A late convert

The problem was partly answered by the first contribution of J. Tuzo Wilson, the University of Toronto's first Professor of Geophysics, a field-based geologist and applied geophysicist. He was a late convert to the concepts beginning to emerge concerning a mobile earth in the late 1950s and with all the insight of the true believer he wrote a remarkable series of reflections on the large-scale consequences of continental drift, including in 1962 a convectional driver for ocean evolution (Fig. 16.2). Three years later came his spectacular interpretation of oceanic fracture zones that cut across mid-ocean ridges – first imaged in the magnificent sea-floor maps of Bruce Heezen, Marie Tharp and Maurice Ewing published in 1959. Hess had already noted such features in 1954 as: 'Fault scarps commonly at high angles to the ridge axis are found on some [ridges] and show relatively high seismicity.'

In his 1965 *Nature* paper, Wilson designated oceanic fracture zones as a new type of shear or strike-slip fault, which he called 'transforms', the name noting their role in changing plate boundaries from extension to shearing or to the reverse thrust faulting seen at oceanic trenches. In his analysis he very cleverly inferred that they

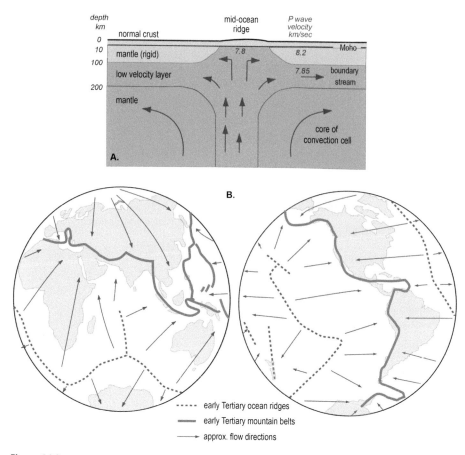

Figure 16.2 A. Version of J. Tuzo Wilson's 1962 section across a mid-ocean ridge 'suggesting one possible underlying structure'. Specifically, the structure involves his identification of Gutenberg's 'low velocity layer (LVL)' as a sliding plane (Wilson's 'boundary stream') for the rigid upper mantle over a deeper ascending convection cell carrying melt up to the Moho – note the similarity of P-wave velocities for the rising melt and the LVL. Otherwise, it is very similar to Holmes's 1944 and Hess's 1962 representations, but without their emphasis on near-surface magma rise, dyke injection and ridge volcanism. B. Wilson's companion figure of 1962 showing a 'World map of continental and mid-ocean mountain systems in early Tertiary time showing the directions in which sub-crustal currents might have moved'.

could only be active between the ridge segments separated by them (Fig. 16.3) and that they disobeyed the rules of conventional wrench faulting – the ridge axis was displaced in a direction opposite to the orientation of the shear couple responsible. His interpretation was amply confirmed two years later by Lynn Sykes, who was able to demonstrate the active and passive parts of the fractures from the presence or absence of seismic activity along their arc-like trajectories.

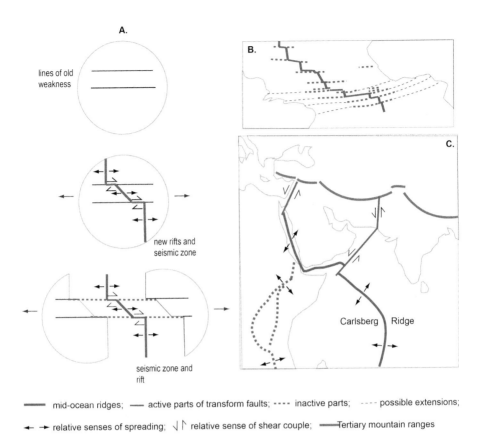

lines of old weakness

new rifts and seismic zone

seismic zone and rift

A.

B.

C.

Carlsberg   Ridge

mid-ocean ridges; ── active parts of transform faults; ‑‑‑‑ inactive parts; ‑‑‑‑ possible extensions;

◄─ ─► relative senses of spreading; √ Γ relative sense of shear couple; ▬Tertiary mountain ranges

Figure 16.3 Versions of the diagrams from Wilson's 1965 paper, captions in his own words. A. Diagram illustrating three stages in the rifting of a continent into two parts (for example, South America and Africa). There will be seismic activity along the heavy [red] lines only. B. Sketch (after Krause and Heezen et al.) showing how the Mid-Atlantic ridge is offset to the left by active transform faults which have dextral [right-handed] motions if the rift is expanding. C. 'Sketch illustrating the end of the Carlsberg mid-ocean ridge by a large transform fault...extending to the Hindu Kush, the end of the rift up the Red Sea by a similar transform fault extending into Turkey and the still younger East African rifts.'

This latter feature gave the opportunity for their use as geometers in determining the pole of rotation that each moving piece of translating plate must possess. So was born the idea of 'tectonics on a sphere' as developed by Dan McKenzie and Robert Parker in 1967. This defined the fundamental notation (from a theorem by the eighteenth-century geometer Leonhard Euler) necessary to give a vectorial analysis of moving, rigid plates over the spherical earth. Their motions and interactions at ocean ridges, trenches and along transform faults enabled what they termed 'pavement tectonics' to occur. The term 'plate' was first used, I think, by Wilson in his 1965 paper but was thoroughly re-christened in a classic paper in *Journal of Geophysical Research* by Bryan Isacks, Jack Oliver and Lynn Sykes, 'Seismology and the new global tectonics' in 1968. Oliver's later 'geolimerick' summarizes the importance of the whole process in a light-hearted but profound way:

> The plates in dynamic mosaic
> Through history both fresh and archaic
> Like bold engineers
> For some two billion years
> Have kept Earth from becoming prosaic

Where have all the oceans gone?

A final development was the application of plate tectonics to the evolution of ancient oceans and their transformation into mountain belts. That such an approach was both possible and desirable came initially from the previously discussed results of workers on ancient ophiolites, and by the 1962 comments of Hess himself when he speculated concerning the youthfulness of the current ocean ridges (he thought only 100-200 million years old) versus the possibility of evidence in the geological record for ancient examples. He wrote:

> The question may be asked: Where are the Palaeozoic and Precambrian mid-ocean ridges, or did the development of such features begin rather recently in the Earth's history?

This question was answered in 1966 by another question from Wilson in his third great *Nature* paper of the 1960s – he asked 'Did the Atlantic Ocean close and then re-open?' His wide reading and field knowledge of the stratigraphy and palaeontology of both northern European and North American Palaeozoic rocks led him to speculate on the likely existence of a Proto-Atlantic Ocean that would substantially explain the very great difference between Cambrian fossil bottom-living faunas of Appalachia and Scotto-Scandinavia compared with those of Maritime Canada, Anglo-Wales and the rest of mainland Europe (Fig. 16.4). His mobilist tendencies, now roused to their peak, enabled him to latch onto A.W. Grabau's initial idea that a former Proto-Atlantic had existed and that it was closed before the opening of the modern Atlantic. Such a sequence of events would explain the fossil distributions. In his own words:

> Four lines of evidence suggest that this proposal is reasonable. (Unfortunately, so far as I can ascertain, palaeomagnetic evidence which might bear on this problem does not exist.)

First, this reconstruction of geological history is held to provide a unified explanation of the changes in rock types, fossils, mountain building episodes and palaeoclimates represented by the rocks of the Atlantic region.

Second, wherever the junction between contiguous parts of different [faunal] realms is exposed, it is marked by extensive faulting, thrusting and crushing.

Third, there is evidence that the junction is everywhere along the eastern side of a series of ancient island arcs.

Fourth, the fit appears to meet the geometric requirement that during a single cycle of closing and reopening of an ocean, and in any latitudinal belt of the ocean, only one of the pair of opposing coasts can change sides.

He proposed that the history of the North Atlantic region involved a Cambrian open ocean that closed by stages in the Ordovician and Silurian. It would take another few years for Jim Briden's group at Leeds to obtain the direct palaeomagnetic evidence

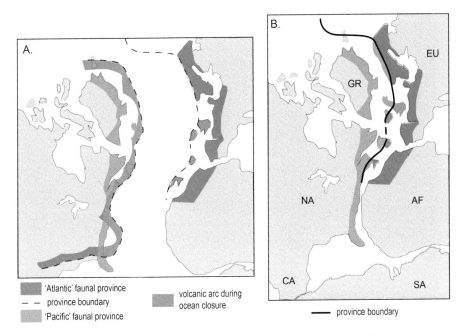

Figure 16.4 Versions of maps that illustrate Wilson's theory of a proto-Atlantic Ocean (subsequently named Iapetus) that separated distinct fossil faunal realms (A) and which must have closed in late-Lower Palaeozoic times (B).

that Wilson had required for the extent of the Proto-Atlantic to be established. Their 1973 paper took results from several years' study of the palaeomagnetism of ophiolitic, arc volcanic and plutonic igneous rocks to constrain any post-early Ordovician width to 1000+/- 800 km. During late Palaeozoic to early Mesozoic times, shallow shelf seas then covered the North Atlantic area, successively invaded by waters of the Tethyan ocean to the south. By the late-Mesozoic and into the Tertiary the modern North Atlantic opened.

The whole issue of extinct oceans raised by Wilson was explored and refined in 1969 when John Dewey published in *Nature* his 'Evolution of the Caledonian/ Appalachian Orogen', a detailed geological reconstruction of the Lower Palaeozoic evolution of the classic ancient mountain belt. In the spirit of the 'new global tectonics' heralded by Isaacs, Oliver and Sykes the previous summer, Dewey regarded the belt as the site of closure of Wilson's Proto-Atlantic. The evidence came not only from the faunal provinces used by Wilson, but also in the form of Ordovician ophiolites of the Scottish Southern Uplands and Highland Border and in Newfoundland. Additional evidence came from volcanic arcs produced by the southwards subduction of the Iapetus plate forming the Ordovician volcanic arc terranes of the Lake District and North Wales, as subsequently demonstrated geochemically in a pioneering paper by J. Godfrey Fitton and D.J. Hughes in 1970. The ocean was splendidly and appropriately named as Iapetus by Brian Harland and Rod Gayer in 1972, after the Greek god, father of Atlas, from whom the Atlantic was originally named.

As an undergraduate on an inter-university geological excursion around Ireland in the summer of 1968 led by John Dewey, with 15 or so other students I was treated to a preview of his *magnum opus* in the resident's lounge of the Abercorn Arms Hotel in Strabane, Co. Tyrone. It came at the very end of the excursion after a late-evening seminar discussion, John explaining with large drafts of maps that were to feature in his paper of the next year. It was an intellectual eye-opener for us all, an occasion that I for one have never forgotten, deep in geological insight provided by the uniformitarianism of modern plate tectonics as used to reconstruct the evolution of an ancient mountain belt and its oceanic and tectonic history.

17

Carbon, mountains and cooling

We left Harold Urey's scientific journey through his long life with the silicate weathering reaction as a crustal buffer that kept a small, stable level of carbon dioxide in the atmosphere, mentioning the fact that he had gloriously re-invented the wheel! Here we pursue the topic historically in the light of today's knowledge that there has been a steady rise, also an increasing rate of rise, in atmospheric carbon dioxide concentration recorded at the Mauna Loa observatory in Hawaii (Fig. 17.1). Over the past sixty years it has increased from 315 to 411 parts per million, around 30%. The likelihood is that this is because of the post-1940s global industrial and transport revolutions with the expansion of fossil fuel burning and related CO_2 emissions.

The earliest and most prescient remark concerning the effects of machine-generated carbon dioxide came from polymath Charles Babbage in 1835:

> The force of vapour [steam] is another fertile source of moving power; but even in this case it cannot be maintained that power is created. Water is converted into elastic vapour by the combustion of fuel. The chemical changes which thus take place are constantly increasing the atmosphere by large quantities of carbonic acid [carbon dioxide] and other gases noxious to animal life. The means by which nature decomposes these elements, or reconverts them into solid form [carbon], are not sufficiently known.

He had warned (in a book, *On the Economy of Machinery and Manufactures*, which extolled the advantages of industrialization!) that there might be, as we would put it today, 'unintended consequences' to widespread coal burning.

A brilliant reply to Babbage's request for further research into the carbon problem came ten years later from another fertile mind, that of the young French geochemist, Jacques-Joseph Ébelmen. He was born in Baume in eastern France and was originally a mining engineer. His early analytical work on the chemical assaying of industrial

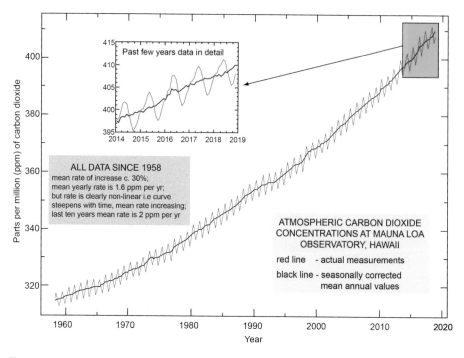

Figure 17.1 The record of measured atmospheric carbon dioxide at the Hawaii Observatory for the past sixty years. Source: https://www.esrl.noaa.gov/gmd/ccgg/trends/full.html

rocks and minerals led to improvements in Sèvres porcelain manufacture. He was fêted by Michael Faraday at the 1851 Great Exhibition in London where he was an official French delegate. The two men talked together, probably mostly about the chemical analysis of ancient and modern pottery glazes, a subject of great mutual interest. Faraday invited him to one of his Royal Institution evening lectures but if they ever did discuss Ébelmen's conclusions concerning the nature of chemical weathering by the atmosphere there is no hint that Faraday ever returned to the topic or referred to it. Ébelmen died the next year from a form of brain fever, only thirty-seven years old.

In his 1845 paper, Ébelmen listed the chief natural and human-induced chemical reactions that might change the proportions of carbon dioxide and oxygen in the atmosphere. The results were published in the scientific journal of the French Bureau of Mines and virtually lost to science, never cited nor even acknowledged more than just once until the mid 1980s, by which time Ébelmen's whole scheme had been re-invented by Urey and subsequently incorporated in what became known as the Carbon Cycle. One is tempted to suggest that Carbon Cycling should be known as Ébelmen Cycling in his memory (one thinks of the Krebs Cycle in biochemistry, for example).

Ébelmen had developed a deep understanding of the chemical weathering of rocks. Like Urey he had identified a major 'sink' (depository) for atmospheric and soil carbon dioxide, pointing out first that limestone weathering by carbonic acid in the atmosphere has no effect whatsoever on the carbon cycle. This is because once drained into the oceans, reactions that precipitate calcium carbonate give off carbon dioxide in the same proportions as that used up during the original weathering. The situation was quite different for the weathering of calcium-bearing silicate minerals in the granitic and basaltic rocks that dominate the earth's crust. Here, for two molecules of carbon dioxide used up in weathering only one is given off by subsequent reactions back to calcium carbonate. The overall carbon dioxide of the atmosphere is thus reduced.

On another tack, Michael Faraday's successor as Director of the Royal Institution, London, John Tyndall, was a hugely talented experimental physicist with wider interests in glaciology, geology and evolutionary theory. He had closely studied Fourier's work on the diffusion of heat by conduction, in particular the idea that the earth's atmosphere might be instrumental in keeping incoming solar heat bound closely to the planetary surface. He decided to experimentally test the ability of various atmospheric gases to absorb heat energy using a novel instrument of his own invention. He identified water vapour, carbon dioxide and methane as the chief absorbers. These were all polyatomic, i.e., comprising more than one type of atom unlike, say, oxygen or nitrogen. In 1859 he published his discovery paper that was to lead to the 'greenhouse' concept:

> The bearing of this experiment upon the action of planetary atmospheres is obvious...the atmosphere admits of the entrance of the solar heat, but checks its exit; and the result is a tendency to accumulate heat at the surface of the planet...gases absorb radiant heat of different qualities in different degrees.

Neither Tyndall nor anyone else at the time considered that human activity could influence the amount of insulating gases like carbon dioxide in the atmosphere. Yet, perversely, the cooling effect of a reduction in carbon dioxide concentration was topical. Only eighteen years before, Louis Agassiz had staggered the scientific world by proposing the existence of past ice ages, and by this time in the late 1850s the geological evidence for this theory was widely accepted.

It was left to the future Swedish Nobel prize winner, Svante Arrhenius, to have a further crack at exploring the possibility of a carbon dioxide effect arising from industrial emissions. In December 1895 he read a landmark paper to the Swedish Academy of Sciences making use of Tyndall's and later researchers' experimental results. He calculated the timing and effect of changes in carbon dioxide concentrations on the atmospheric balance that might cause global cooling or warming. An extract from this lecture was published in English the next year. Strangely, he never considered a human (industrial) source for carbon dioxide, rather he was concerned with changing carbon dioxide concentrations in general, from natural causes not specified, and how they might be responsible. As he writes, rather wearily, in his summary discussion:

> I should certainly not have undertaken these tedious calculations if an extraordinary interest had not been connected with them...on the probable causes of the Ice Age; and...that there exists as yet no satisfactory hypothesis that could explain how the climatic conditions for an ice age could be realized in so short a time as that which has elapsed from the days of the glacial epoch.

As it turns out, it seems Arrhenius was misled by the faulty conclusions of a close colleague. In his summary discussion, he gives lengthy quotes from geologist Arvid Högbom who, clearly unaware of the rates of weathering reactions, had concluded that the:

> quantity of carbonic acid, which is supplied to the atmosphere chiefly by modern industry, may be regarded as completely compensating the quantity of carbonic acid that is consumed in the formation of limestone (or other mineral carbonates) by the weathering or decomposition of silicates.

And that:

> the most important of all the processes by means of which carbonic acid has been removed from the atmosphere in all times, namely the chemical weathering of siliceous minerals, is of the same order of magnitude as a process of contrary effect, which is caused by the industrial development of our time, and which must be conceived as being of a temporary nature.

In other words, there was nothing for humans to worry about. But, vitally, Högbom was wrong about the equivalence of rates of the human-induced supply of CO_2 and its removal by silicate weathering. The latter is a slow process by comparison – any buffering effect by weathering cannot begin to influence the volumes of human-induced outpourings of carbon dioxide on the timescales involved. Högbom considered the ups and downs of carbon dioxide concentrations that might have occurred in the geological past as wholly random – due to variations in volcanic eruption intensities. Arrhenius and others by this time were calling the atmospheric warming due to carbon dioxide the 'hot-house effect'.

The cycling of the solid earth we recognize today comes courtesy of plate tectonics. At the same time as these tectonic concepts were evolving there were attempts from the early 1970s to revisit Hutton and Urey's visions of global cycling in what became known as 'biogeochemical cycling'. This describes gaseous and aqueous reactions during volcanic eruptions, chemical weathering and photosynthesis. Such approaches bore fruit when applied to sedimentary cycling and laid the basis for 'hindcast' modelling (predicting the consequences of what is already there and known), in particular that of the geological carbon cycle (Fig. 17.2) deduced by Robert Berner and co-workers.

Cycling concepts led to several questions concerning the nature of the young Earth (i.e., pre-2500 Ma) envisioned by Urey and Miller, notably the timing and mechanism for the appearance of an oxygenated atmosphere. That an atmosphere had evolved from reducing to oxidizing was ascribed to the cumulative effects of bacterial and early-plant photosynthesis and carbon burial. Such ideas and lateral thinking from research on planetary atmospheres led to the concept of Earth as a self-regulating entity, the state of Gaia, as proposed in 1979 in a widely read book by James Lovelock.

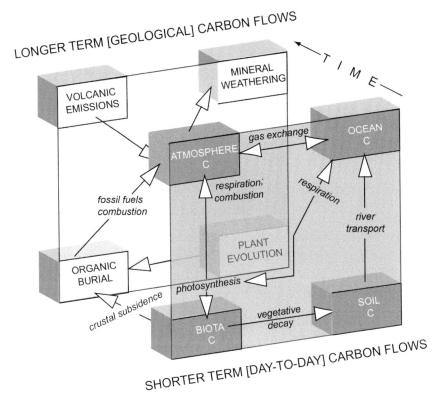

Figure 17.2 Three-dimensional version of a diagram to illustrate carbon cycles and flows composed after a two-dimensional original by R. Berner (2004). The entire carbon cycle is envisaged as a group of interconnecting boxes representing the major reservoirs of solid, dissolved and gaseous carbon. The arrows with words represent the flow of carbon from one box to another. The total amounts of carbon in these boxes varies – the size of boxes is not proportional to the actual amounts of carbon dioxide. Incoming arrows to a box represent additions of carbon dioxide; outgoing arrows from a box represent subtractions.

When first proposed, this scheme operated without reference to, and entirely independently of, the by then well-known concepts of deep time and plate tectonics – they seemed fatal omissions to this earth scientist as he disappointedly read through the book in Leeds in 1980.

The essential point that Lovelock and others failed to appreciate was that in nature, sedimentation and crustal subsidence will preserve carbon and other sedimentary-entombed elements until they are released to be oxidized by tectonic uplift and erosion. Therefore, elemental burial must be seen in the time context of plate-tectonic recycling, i.e. over several tens to hundreds of million years – a point tellingly raised by J.M. Robinson in her fine paper of 1991.

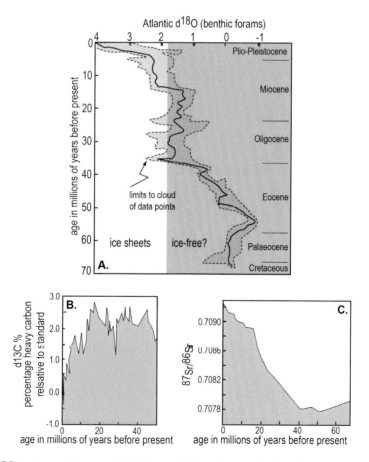

Figure 17.3 Versions of Raymo and Ruddiman 1992. A. The usually slow, but sometimes rapid, change towards heavier oxygen stable isotopes in benthic microfossil calcium carbonate shells during the Tertiary period down to the present day. This is interpreted to be due to extreme oceanic temperature changes consequent on the development of polar ice sheets from the late Eocene onwards and their expansion in the Plio-Pleistocene. B/C. Corresponding changes in strontium isotopes over the same period.

It can be argued that the combination of tectonic and biogeochemical cycling is the great fulfilment of the Huttonian circulation scheme. It finally came about with the realization by Maureen Raymo and William Ruddiman in the 1990s of the implications of Ébelmen-Urey type carbon sequestration in the newly formed Alpine–Himalayan mountain belt around 40–45 million years ago. These new mountains, stretching from the Pyrenees to Sichuan, were full of pristine silicate-bearing rock just waiting to slowly draw down carbon dioxide from the atmosphere as they and their deposits were weathered by warm monsoon rains. This realization led the authors to propose that Tertiary tectonic uplift was the likely cause of the undeniable global cooling seen since the mid-Cenozoic – ending in the Pleistocene Ice Ages (Fig. 17.3).

In support of this theory, enhanced silicate weathering had entirely independent consequences. The young crustal rocks in the Alpine–Himalayan mountains produced by crustal melting and fusion have, in parts, relatively high concentrations of strontium-87, an isotope of the radioactive element rubidium, compared to the more abundant form, strontium-86. Over Mid-Cenozoic to Holocene times the oceans received increasing amounts of strontium-87 relative to strontium-86, as recorded in fossilized calcite and phosphatic shell-forming organisms collected from deep-sea cores.

We therefore see a global outcome – silicate weathering joining plate tectonics as a major environmental control on climate. This realization opened the search for other such linkages in the geological past. In particular it seems possible that the global cooling responsible for the great Permo-Carboniferous glaciations of the Gondwanan continents may have been triggered by the weathering of the Variscan mountain chain. Given the efficacy of the greenhouse effect and the knowledge that the geological carbon cycle interacts with tectonics over long timescales, the causes of past global climate changes need to be carefully re-assessed.

18

Isotopes, ice and orbits

We return to the role of planetary orbits and their effect on the sun's radiation to the upper atmosphere over geological time. As discussed in Chapter 12, an endgame had been reached by Milanković's final mathematical workings for the insolation curves in the late 1930s. At that time there was no chronology available that could test these predictions, and this absence proved crucial in the reticence of many to accept Milanković–Croll orbital theory. All this changed with oceanographic explorations and technical developments from the 1960s onwards. As with the discovery of sea-floor spreading, it was the oceans that were to provide the necessary evidence from refined shipboard and laboratory instrumentation.

Tiny creatures record oceanic goings-on

Holocene and late-Pleistocene palaeontology (also the science of archaeology) had been revolutionized by Willard Libby's post-war development of radiocarbon dating, later refined by the development of accelerator mass spectrometry that enabled organic and organically secreted materials containing minute quantities of ^{14}C to be accurately dated up to around 45 thousand years ago (45 ka). Developments in ocean-floor sampling had moved on from tarred ropes, dredges and grab samples to piston coring, enabling the capture of long, undisturbed bottom sediment cores. The thicker cores, up to twenty metres long, were taken in their hundreds over the world's oceans by survey ships from the chief maritime nations. Such undisturbed cores could be sampled for the microfossil remains of creatures who had lived in contrasting habitats – benthic (near-bottom) and pelagic (water column) invertebrates and plants.

Most important amongst the microfossil creatures were the unicellular calcareous foraminifera ('forams'), which carried the imprint of contemporary sea water temperature in the oxygen stable isotope composition of their tiny and fragile calcium carbonate shells. Using these, a pioneering series of papers over the years from 1955 to 1966 by Cesare Emiliani, a doctoral student of Harold Urey, had established that there had been an isotopic response to the latest (Pleistocene–Holocene) glacial cycles in the oceans. Emiliani reasoned that this response was caused by fluctuating ocean-water temperatures from more frigid glacials to warmer interglacials.

By 1965, the mass spectrometers used to measure the relative proportions of the stable isotopes of oxygen had been so refined that precise measurements could be gained from very small, hand-picked samples taken from precise depths in sediment cores. Chief among the geoscientists who developed such instrumental improvements (modifications to the gas inlet and recording systems) in the mid-1960s was Nicholas Shackleton of the Department of Quaternary Research at the University of Cambridge, England. These enabled him to more rapidly process his samples and to obtain precise temperature estimates from their oxygen stable isotopes. As he reported in 1965 whilst still a doctoral student at Cambridge:

> Modifications are described which enable a commercially available British mass spectrometer to measure small differences in oxygen and carbon isotope ratios with a precision of ±0.01% and using as little as 0.1 ml carbon dioxide sample. This precision is needed for geological work but might also find application in the use of stable isotopes…in biological systems…

The 'geological work' mentioned here, part of his Ph.D. studies, took the form of an audacious test of Emiliani's conclusions regarding ocean-water temperature changes. His results were published in *Nature* in 1967 as 'Oxygen isotope analyses and Pleistocene temperatures re-assessed'. The impetus for this now-classic paper was equally original work published by Eric Olausson two years earlier of an alternative interpretation to Emiliani's sea-water temperature idea. Olausson reasoned that the effect of sea-water fractionation due to temperature on oxygen isotopic values was a small effect but that high latitude glacial ice had formed from water vapour that had undergone major fractionation on its poleward journey from the tropics and that the oceans must therefore have been more significantly depleted in the lighter ^{16}O. So, by calculating the likely volume of the last Ice Age glacial accumulations and expressing these as a fraction of total ocean-water volume, Olausson was able to suggest that there would have been a significant difference in global seawater isotopic composition towards heavier oxygen during the Ice Age glacial maximum compared to that of the Holocene interglacial minimum.

This 'Olausson effect' was well within the limits for detection by Shackleton's mass spectrometer and, after checking Olausson's calculated ice volumes for himself, he reasoned that a direct test of the results would come from a selection of benthic forams to analyse isotopically:

> If such large changes in ocean isotopic composition have taken place, analyses of benthic foraminifera, living in an environment of rather constant temperature, should yield exactly the same record as do analyses of planktonic foraminifera.

So, picking out such forams from samples taken from the same Caribbean core as Emiliani had used to analyse the planktonic record, Shackleton was able to show that such large compositional changes were indeed seen in the benthonic record, though with some inevitable scatter of sample points (Fig. 18.1). After a further overview of the literature, he was able to conclude:

> We therefore have clear evidence that Atlantic and Caribbean surface waters, Atlantic bottom water, Pacific bottom water, and Atlantic deep water as represented by Caribbean bottom water, all registered a glacial–interglacial change in isotopic composition of about 1.4–1.6 parts per thousand, as predicted by realistic calculations.

In a generous gesture towards Emiliani, Shackleton ended his paper by beckoning in a new era, not only for oceanography, but also palaeoclimate and tectonics:

> Although it has generally been assumed that the temperature changes recorded in deep-sea cores were synchronous with the principal glacial episodes in the Northern hemisphere, this could only be proved for those glacial episodes which can be dated radiometrically. On the other hand, if the faunal and isotopic changes which have been observed are the result of the extraction of large quantities of water from the oceans and their storage in the form of ice, their relation to glacial events is unquestionable. It is simply necessary that every faunal or isotopic curve be re-read, taking 'cold' to mean 'extensive continental glaciation' and 'warm' to mean 'glaciers reduced to their present level'.

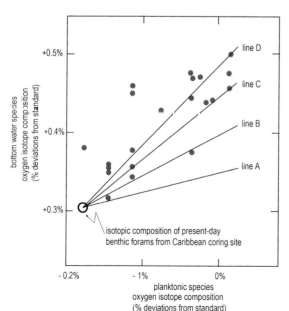

Figure 18.1 Version of Shackleton's 1967 demonstration that benthic and planktonic forams of the same age show oxygen isotope fractionation far more than that expected for sea-water temperature change during glacial periods as indicated by lines A–C. Instead, their fractionations cluster around line D, indicating a greater oceanic compositional change in the isotopes due to freshwater light oxygen extraction and storage in vast Pleistocene ice sheets.

In conclusion, it should therefore be emphasized that the time sequence which Emiliani has been able to obtain by the analysis and correlation of many deep-sea cores remains of inestimable value; indeed, its value is in a sense enhanced by the certainty that it is a time-sequence for terrestrial glacial events, rather than oceanographic events. His work remains a uniquely valuable contribution to geology and equally to archaeology.

Written in water

Shackleton had not only turned the ocean floor sedimentary record into a memory tape for ancient climate change but, *ipso facto*, a potential vital key in the evaluation of Milankovič's orbital control on palaeoclimate: a staggering result. Yet there was a major missing link in the data – this was an independent timescale that might be used to date the various ups and downs of climate predicted by orbital theory. This link was to be supplied by traditional fossil evidence and by palaeomagnetic reversals, then, by way of judicious tuning, by determination of the timing of the orbital parameters themselves.

Magnetic stratigraphy involves the use of radiometric ages to determine the beginning and end of the periods of reversed magnetization that had so fascinated Patrick Blackett (Chapter 6). Since such reversals are global and seemingly quite rapid, they revolutionized the study of Quaternary and older stratigraphic sequences. The first oceanic results were published in the *Bulletin of the Geological Society of America* for 1969 in a paper titled 'Pliocene-Pleistocene sediments of the equatorial Pacific: their paleomagnetic, biostratigraphic, and climatic record'. The authors, J.D. Hays. T. Saito, N.D. Opdyke and L.H. Burckle, were an interdisciplinary team who established an astonishing story of glacial cyclicity extending way back into the Pleistocene using data gained from forams and from palaeomagnetic results on samples taken from thick piston cores. The authors established a 4.4-million-year record back to the Gilbert magnetic epoch. An interval in our current Bruhnes epoch contained eight distinct cycles of fluctuating calcium carbonate content that could be attributed to eight successive glacial/interglacial alternations over the past 700 ky.

But more was to come, for Shackleton then teamed up with Neil Opdyke to produce the most detailed and precise delineation of glacial–interglacial cycles using a combination of palaeomagnetic and stable isotope data gained from two thick (16 and 21 metres) piston cores taken from the equatorial Pacific over the Solomon Plateau. In their paper of 1973 in the journal *Quaternary Research*, the boundaries of 22 glacial/interglacial cycle boundaries due to high and low Northern Hemisphere ice volume were recognized and dated back as far as 870 ka, the longest record yet of changing oceanic composition (Fig. 18.2A). Their paper also enabled a close comparison between benthic and pelagic foram oxygen isotope fluctuations in the most recent glacial cycle – their close match elegantly confirming Shackleton's by now

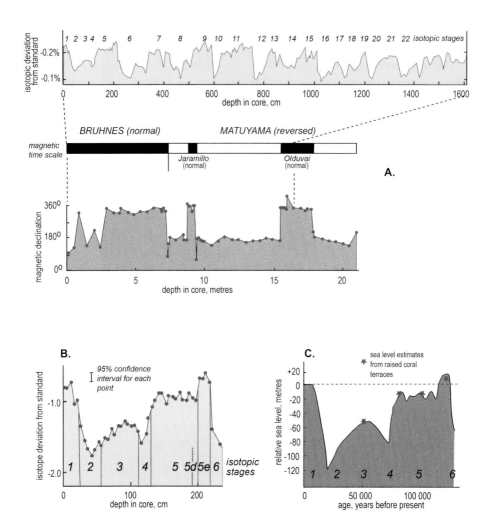

Figure 18.2 Versions of summary diagrams and graphs from Shackleton and Opdyke's 1973 paper. A. Magnetic polarity signatures, normal and reversed, from their deepest core, together with oxygen isotope signatures gained from mass spectrometry analysis of planktonic forams. This record goes back almost 900 ka. B. Detailed isotopic record of the top part of the shallowest core showing the exact changes over the last glacial and interglacial cycles (see text). C. The astonishing, first-ever record of sea-level variation for the last 130 ka gained from calculations concerning the degree of global seawater isotope fractionation given by the results inherent in graph B. Note the close correspondence between these and prior estimates from uplifted coral terraces in Barbados and New Guinea.

widely accepted hypothesis that Quaternary global isotopic changes were due entirely to fluctuating ice volumes.

Further advances with profound future applications came from this paper. First was the use of the isotopic record for the past 130 ky to determine sea-level variations (Fig. 18.2B-C). The authors reckoned that a 0.01% increase in the heavier ^{18}O would result from each 10-metre sea-level drop down to the total fall of 120 metres at the last glacial maximum (LGM) some 17 000 ka. The resulting sea-level curve shows the precipitous rapid rise after the LGM into our modern Holocene interglacial to the even higher (by 6 or so metres) mean sea level recorded during the peak of the long last interglacial around 130 ka. Here is where the story begins to sound convincing, since, as Shackleton and Opdyke pointed out, there were some independent checks on their predictions:

> We also plot…the estimated position of sea-level maxima at about 120 000, 100 000 and 80 000 years ago deduced by Broecker et al….on the basis of work [from uplifted coral terraces dated radiometrically] on Barbados, and about 50 000 years deduced by Veeh and Chappell on the basis of similar work in New Guinea…The satisfactory agreement gives strong support for our contention that ice volume, or sea level, may be reliably estimated from the isotope data. Important continuing work in Barbados and New Guinea…will permit a more accurate conversion factor between isotopic composition and sea level to be determined, using other calibration points in addition to those mentioned relating to the past 120 000 years. This in turn will enable us to use the isotopic variations in the earlier part of the Pleistocene as a standard sea level curve against which to calibrate long-term uplift rates in these same areas.

The authors concluded that since the relative positions of sea level were '…derived by two entirely independent lines of reasoning [they] must quell any remaining doubt as to the correctness of this correlation.' Later work on these matters has changed some of the dates and rates by small amounts, up to 10%, but the main features of this first global sea-level curve stand robustly and have given much information, as the authors predicted, on the rates and distribution of tectonic uplift over many parts of the seismically active earth's surface in dated sea-level deposits, from coastal California to the Gulf of Corinth rift and elsewhere.

Second was the extension of the marine record that the authors could make over the full length of their longest core (Fig. 18.2A). Over a thickness of 16 metres sampled at 10 cm intervals, the calcareous oozes yielded the most perfect worldwide subdivisions of the Pleistocene period ever proposed, one based firmly on the fixed volumes of stored and melted ice over eleven or more glacial cycles. The authors proudly wrote:

Thus it is highly unlikely that any superior stratigraphic subdivision of the Pleistocene will ever emerge…We propose that the stages set up in this core be adopted as standard for the latter half of the Pleistocene.

Orbital theory vindicated statistically

After publication of the 1973 paper, Shackleton joined forces with John Hays to work with John Imbrie on a statistical analysis of their oceanic oxygen isotope records.

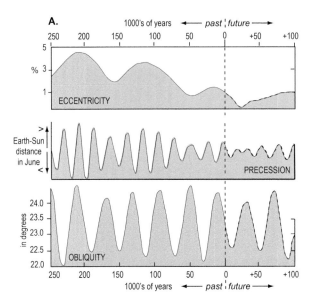

Figure 18.3 A. Calculated changes in eccentricity, precession and tilt over the past 250 ka (data of A. Berger, 1977). B. Spectra of climatic variation over the past half-million years taken from the oxygen isotope data of Figure 19.2A and other sources, revealing the relative importance of the three controlling orbital variations in A, with the eccentricity component out on top. (Hays et al., 1976).

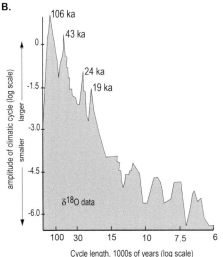

Choosing cores containing relatively rapidly deposited deep-ocean sediments, Imbrie used spectral analysis to try to determine whether the various 'wiggles' appearing on these curves were random in their timing or whether there were distinct time intervals coming in from the eccentricity, tilt and precession orbital factors. The eccentricity effect at *c.*100 ky intervals was already obvious even to the untutored eye, but the other two were not so obvious. The spectral analysis (Fig. 18.3) confirmed these as significant components at 41 and 23 ky intervals. Furthermore, application of a filtering technique enabled Imbrie to show that there was also a several thousand-year lag in the response time of the ice sheets to the beginnings of each tilt and precessional cycle. This enabled him and Kathleen Palmer Imbrie to later write:

> Hays, Imbrie and Shackleton announced their findings in an article in *Science*, which appeared on December 10, 1976: 'Variations in the Earth's orbit: pacemaker of the Ice Ages.' A century after Croll published his theory and 50 years after Milankovitch mailed his radiation curves to Köppen and Wegener, two cores from the Indian Ocean confirmed the astronomical theory of the ice ages.

James Croll would have been fascinated and delighted. A suitable note, I feel, on which to end this book.

Bibliography by chapters

Aims and Forewords

Bailey, Sir E.B. (1952) *Geological Survey of Great Britain*. Murby, London
Park, G. (2020) *Breakthroughs in Geology: Ideas that Transformed Earth Science*. Dunedin Academic Press, Edinburgh & London
Simmons, M. (2018) *Great Geologists*. Halliburton, Abingdon

Part 1: Deep Stuff

Chapter 1 Richard Dixon Oldham

Bilham, R. and England, P. (2001) Plateau pop-up during the great 1897 Assam earthquake. *Nature* **410**, 806–809
Davison, C. (1936) Richard Dixon Oldham, 1858–1936. *Obituary Notices of Fellows of the Royal Society* **2** (5), 111–113
Mallet, R. (1847) *The Dynamics of Earthquakes*. Proceedings of the Royal Irish Academy **21**
Mallet, R. (1862) *Great Neapolitan Earthquake of 1857*. Chapman and Hall, London
Oldham, R.D. (1899) Report of the great earthquake of 12th June, 1897. *Memoir of the Geological Survey of India* **29**, 1–379
Oldham, R.D. (1900) On the propagation of earthquake motion to great distances. *Philosophical Transactions of the Royal Society* **194A**, 135–174
Oldham, R.D. (1906) The constitution of the Earth as revealed by earthquakes. *Quarterly Journal of the Geological Society of London* **62**, 456–475
Oldham, R.D. (1919) The interior of the Earth. *Geological Magazine* **6**/1, 18–27

Chapter 2 Andrija Mohorovičić

Herak, M. (2005) *Andrija Mohorovičić's memorial rooms: Mohorovičić Discontinuity*. http://www.gfz.hr/sobe-en/discontinuity.htm
Herak, D. and Herak, M. (2010) The Kupa Valley (Croatia) earthquake of 8 October 1909 – 100 Years Later. *Seismological Research Letters* 81(**1**), 30–36

Chapter 3 Beno Gutenberg

Barrell, J. (1914) The strength of the earth's crust. *Journal of Geology* **22**, 655–683
Gutenberg, B. (1926) Untersuchungen zur Frage, bis zu welcher Tiefe die Tirde kristallinist. *Zeitschrift Geophysics* **2**, 24–29

Gutenberg, B. (1936) Structure of the earth's crust and the spreading of the continents. *Geological Society of America Bulletin* **47**/10, 1587–1610

Gutenberg, B. (1948) On the layer of relatively low wave velocity at a depth of about 80 kilometres. *Bulletin of the Seismological Society of America* **38**, 121–148

Gutenberg, B. (1952) Waves from blasts recorded in Southern California. *Transactions of the American Geophysical Union* **43**, 223–232

Gutenberg, B. (1955) Channel waves in the Earth's crust. *Geophysics* **20**, 283–294

Gutenberg, B. (1959a) The asthenosphere low velocity layer. *Annals of Geophysics* **12**(4), 439–460

Gutenberg, B. (1959b) *Physics of the Earth's Interior*. Academic Press, New York

Gutenberg, B. (1960) Low-velocity layers in the earth, ocean and atmosphere. *Science* **131**, 959–965

Gutenberg, B., and Richter, C. F. (1939) New evidence for a change in physical conditions at depths near 100 kilometres. *Bulletin of the Seismological Society of America* **29**, 531–539

Gutenberg, H. (1981) Interview by Mary Terrall. Pasadena, California, 6 and 13 February 1980. *Oral History Project, California Institute of Technology Archives* http://resolver.caltech.edu/CaltechOH:OH_Gutenberg_H

Knopoff, L. (1999) Beno Gutenberg. *Biographical Memoirs of the National Academy of Sciences* **76**, 115–147

Chapter 4 Inge Lehmann

Bolt, B.A. (1987) 50 years of studies on the inner core. *EOS* **68**/6

Bolt, B.A. (1997) Inge Lehmann, 13 May 1888–21; Elected Foreign Member of the Royal Society, 1969. *Biographical Memoirs of Fellows of the Royal Society*

Bolt, B.A. and Hjortenberg, E. (1994) Memorial Essay, Inge Lehmann (1888–1993). *Bulletin of the Seismological Society of America* **84**, 229–233

Kölbl-Ebert, M. (2001) Inge Lehmann's paper: P′ (1936). *Classic Papers in the History of Geology, Episodes* **24**, 262–267

Lehmann, I. (1936) P′. *Bureau Central Séismologique International Strasbourg: Publications du Bureau Central Scientifiques* **14**, 87–115

Lehmann, I. (1987) Seismology in the Days of Old. *EOS*, **68**/3

Part 2: Drifting Stuff

Chapter 5 Alfred Wegener

Georgi, J. (1962) Memories of Alfred Wegener. *International Geophysics* **3**, 309–324

Greene, M.T. (2015) *Alfred Wegener: Science, Exploration, and the Theory of Continental Drift*. Johns Hopkins University Press, Baltimore

Jacoby, W.R. (1981) Modern concepts of Earth dynamics anticipated by Alfred Wegener in 1912. *Geology* **9**, 25–27

Jacoby, W.R. (2001) Translation of *Die Entstehung der Kontinente* Dr Alfred Wegener Petermanns Geographische Mitteilungen, 58 I, 185–195, 253–256, 305–309, 1912. *Journal of Geodynamics* **32**, 29–63

Jacoby, W.R. (2012) Alfred Wegener: 100 Years of Mobilism. *Geoscientist* **22**/9, 13–17

Köppen, W. and Wegener, A. (1924) *Die Klimate der geologischen Vorzeit*. Berlin. English translation and facsimile (2015) *The Climates of the Geological Past* Borntraeger Science Publishers, transl. B. Oelkers

Oreskes, N. (1999) *The Rejection of Continental Drift: Theory and Method in American Earth Science*. Oxford University Press, New York & Oxford

Summerhayes, C. (2015) Book Review of Köppen, W. and Wegener, A. 1924 Die Klimate der geologischen Vorzeit. *Geoscientist*, November 2015

Wegener, A. (1922) *The Origin of Continents and Oceans*. 3rd edn transl. Skirl, J.G.A. Methuen, London, 1924

Wegener, A. (1928) *The Origin of Continents and Oceans.* 4th edn transl. Biram, J., Dover Publications, 1966

Wegener, E. (1960) *Alfred Wegener.* F.A. Brockhaus, Wiesbaden

Chapter 6 Patrick Maynard Stuart Blackett

Blackett, P.M.S. (1952) A negative experiment relating to magnetism and the earth's rotation. *Philosophical Transactions of the Royal Society of London* **245A**, 303–370

Blackett, P.M.S. (1956) *Lectures on Rock Magnetism.* Weizman Science Press of Israel, Jerusalem

Briden, J.C., Morris, W.A. and Piper, J.D.A. (1973) Palaeomagnetic studies in the British Caledonides – VI. Regional and global implications. *Geophysical Journal of the Royal Astronomical Society*, **34**, 107–134

Bullard, E.C. (1948) The magnetic field within the earth. *Philosophical Transactions of the Royal Society of London* **197**A, 433–453

Bullard, E. (1974) Patrick Blackett: an appreciation. *Nature* **250**, 370

Collinson, D.W., Creer, K.M., Irving, E. and Runcorn, S.K. (1950) The measurement of the permanent magnetization of rocks. *Philosophical Transactions of the Royal Society of London* **250**A, 73–82

Creer, K.M. (1965) Palaeomagnetic data from the Gondwanic continents. In: *A Symposium on Continental Drift*, P.M.S. Blackett, E. Bullard and S.K. Runcorn (eds), Philosophical Transactions of the Royal Society of London A**258**, 27–90

Irving, E. (1956) Palaeomagnetic and palaeoclimatological aspects of polar wandering. *Geofisica Pura e Applicata* **33**, 23–41

Keary, P., Klepeis, K.A. and Vine, F.J. (2009) *Global Tectonics.* Wiley-Blackwell

McElhinny, M.W. and McFadden, Philip L. (2000) *Palaeomagnetism: continents and oceans.* Academic Press, San Diego

Runcorn, S.K. (1956) Palaeomagnetic comparisons between Europe and North America. *Proceedings of the Geological Association of Canada* **8**, 77–85

Runcorn, S.K. (1962) Towards a theory of continental drift. *Nature* **193**, 311–314

Part: 3 Hot Stuff

Chapter 7 Arthur Holmes

Boltwood, B.B. (1907) The ultimate disintegration products of radioactive elements, Part II. The disintegration products of uranium. *American Journal of Science* (Series 4) **23**, 77–88

Eve, A.S. (1939) *Rutherford: Being the Life and Letters of the Rt. Hon. Lord Rutherford.* MacMillan, London

Holmes, A. (1913) *The Age of the Earth.* Harper and Brothers, London & New York

Holmes, A. (1931) Radioactivity and earth movements. *Transactions of the Geological Society of Glasgow* **18**, 559–606

Holmes, A. (1944) *Principles of Physical Geology.* Thomas Nelson and Sons, London; 2nd Edn (1965)

Joly, J. (1899) An estimate of the geological age of the Earth. *Scientific Transactions of the Royal Dublin Society* **7**, 23–66

Kovarik, A. (1929) Biographical Memoir of Bertram Borden Boltwood, 1870–1927. *National Academy of Sciences of the United States of America*, **14**/3

Lewis, C. (2000) *The Dating Game: One Man's Search for the Age of the Earth.* Cambridge University Press

Strutt, R.J. (1904) A study of the radio-activity of certain minerals and mineral waters. *Proceedings of the Royal Society A*, **73**, 191

Strutt, R.J. (1906) On the distribution of radium in the earth's crust, and on the earth's internal heat. *Proceedings of the Royal Society A*, **77**, 472

Chapter 8 Norman Levi Bowen

Bowen, N.L. (1912) The binary system: Na2Al2Si2O8 (nephelite; carneigieite) – CaAl2Si2O8 (anorthite). *American Journal of Science.* 4th Series, **33**, 551–573

Bowen, N.L. (1913) The melting phenomena of the plagioclase feldspars. *American Journal of Science.* 4th Series, **37**, 487–500

Bowen, N.L. (1928) *The Evolution of Igneous Rocks.* Princeton University Press, Princeton, New Jersey. 2nd edn Dover Publications, 1956

Bowen, N.L. and Schairer, J.F. (1935) The system $MgO\text{-}FeO\text{-}SiO_2$. *American Journal of Science.* 5th Series, **29**, 151–217

Daly, R.A. (1926) *Our Mobile Earth.* Charles Scribner's Sons, New York & London

Eugster, H.P. (1980) Norman Levi Bowen. 1887–1956. A biographical memoir. *National Academy of Sciences.* Washington D.C.

Gill, R. (2015) *Chemical Foundations of Geology and Environmental Geoscience.* Wiley-Blackwell, Chichester

Hall, J. (1790) Observations on the formation of granite. *Transactions of the Royal Society of Edinburgh* **3**, 8–12

Hall, J. (1798) Experiments on whinstone and lava. *Transactions of the Royal Society of Edinburgh* **5**, 43–75

Hall, J. (1812) Account of a series of experiments, shewing the effects of compression in modifying the action of heat. *Transactions of the Royal Society of Edinburgh* **6**, 71–184

Tilley, C.E. (1957) Norman Levi Bowen. *Biographical Memoirs of Fellows of the Royal Society*, 7–22

Wyllie, P. J. (1999) Hot little crucibles are pressured to reveal and calibrate igneous processes. In: *James Hutton – Present and Future*, G.Y. Craig & J.H. Hull (eds), Geological Society, London, Special Publication **150**, 37–57

Yoder, H.S., Jr. (1976) *Generation of Basaltic Magma.* National Academy of Sciences, Washington D.C.

Yoder, H.S. Jr. (1998) Norman L. Bowen: The experimental approach to petrology. *GSA Today* May 1998, 10–11

Yoder, H.S. Jr. (1992) Norman L. Bowen (1887–1956), MIT class of 1912, First predoctoral fellow of the Geophysical Laboratory. *Earth Sciences History* **11**, 45–55

Part 4: Stressful Stuff

Chapter 9 Ernest Masson Anderson

Anderson, E.M. (1905) The dynamics of faulting. *Transactions of the Edinburgh Geological Society* **8**, 387–402

Anderson, E.M. (1923) The geology of the schists of the Schiehallion district, Perthshire. *Quarterly Journal of the Geological Society of London, 79, 423–442*

Anderson, E. M. (1936) The dynamics of formation of cone sheets, ring-dykes, and cauldron subsidence. *Proceedings of the Royal Society of Edinburgh* **61**, 128–157

Anderson, E.M. (1937) Cone-sheets and ring-dykes: the dynamical explanation. *Bulletin Volcanologique* **1**, 35–40

Anderson, E.M. (1942) *The Dynamics of Faulting and Dyke Formation with Application to Britain.* (2nd Edn revd 1951) Oliver & Boyd, Edinburgh

Anderson, E.M. (1942) On lineation and petrofabric structure and the shearing movement by which they have been produced. *Quarterly Journal of the Geological Society of London*, **104**, 99–126

Bailey, E.B. and Anderson, E.M. (1925) *The Geology of Staffa, Iona and western Mull.* Memoir of the Geological Survey of Great Britain, Sheet 43 (Scotland)

Barber, A.J. (2010) Peach and Horne: The British Association excursion to Assynt September 1912. In: *Continental Tectonics and Mountain Building: The Legacy of Peach and Horne*, R.D. Law and 4 others (eds), 29–50, Geological Society of London Special Publication **335**

Coward, M.P. (1980) The analysis of flow profiles in a basaltic dyke using strained vesicles. *Journal of the Geological Society of London* **137**, 605–615

Healy, D., Butler, R.W.H., Shipton, Z.K. & Sibson, R.H. (eds) (2012) *Faulting, Fracturing and Igneous Intrusion in the Earth's Crust.* Geological Society of London, Special Publication **367**

Kennedy, W.Q. and Anderson, E.M. (1938) Crustal layers and the origin of magmas. *Bulletin Volcanologique*, 2nd Series, **3**, 23–82

National Health Service and Glasgow City Archives
https://archiveshub.jisc.ac.uk/data/gb812-hb13/hb13/5/191/14

Richey, J.E. (1935) *British Regional Geology: Scotland; the Tertiary Volcanic Districts.* Geological Survey of Great Britain, London

Richey, J.E. (1939) The dykes of Scotland. *Transactions of the Edinburgh Geological Society* **13**, 393–435

Chapter 10 Karl Anton Terzaghi

Casagrande, A. (1960) Karl Terzaghi – his life and achievements. In: *From Theory to Practice in Soil Mechanics: Selections From the Writings of Karl Terzaghi, with Bibliography and Contributions on His Life and Achievements*, Bjerrum, L. et al. (eds), 3–21. John Wiley and Sons, New York & London

Guerriero, V. and Mazzoli, S. (2021) Theory of effective stress in soil and rock and implications for fracturing processes: a review. *Geosciences* **11**, 119 https://doi.org/10.3390/ geosciences11030119

Reynolds, O. (1885) On the dilatancy of media composed of rigid particles in contact, with experimental illustrations. *Philosophical Magazine*, Series 5, **20**, 127

Skempton, A.W. (1953) Soil mechanics in relation to geology. *Proceedings of the Yorkshire Geological Society* **29**, 33–62

Skempton, A.W. (1960) Significance of Terzaghi's concept of effective stress. In: *From Theory to Practice in Soil Mechanics: Selections From the Writings of Karl Terzaghi, with Bibliography and Contributions on His Life and Achievements*, Bjerrum, L. et al. (eds), 42–54. John Wiley and Sons, New York & London

Terzaghi, K. (1943) *Theoretical Soil Mechanics.* John Wiley and Sons, New York

Terzaghi, K., Peck, R.B. and Mesri, G. (1996) *Soil Mechanics in Engineering Practice.* 3rd edn John Wiley and Sons, New York

Chapter 11 Ralph Alger Bagnold

Bagnold, R.A. (1935) The movement of desert sand. *Proceedings of the Royal Society of London* **A163**, 594–620

Bagnold, R.A. (1935) *Libyan Sands: Travels in a Dead World.* Hodder and Stoughton, London; reprinted by Eland Publishing Ltd. (2010)

Bagnold, R.A. (1941) *The Physics of Blown Sand and Desert Dunes.* William Morrow and Company, New York; republished by Chapman and Hall, London (1971)

Bagnold, R.A. (1954) Experiments on a gravity-free dispersion of large solid spheres in a Newtonian fluid under shear. *Proceedings of the Royal Society of London* A**225**, 49–63

Bagnold, R.A. (1966) An approach to the sediment transport problem from general physics. *U.S. Geological Survey Professional Paper* **422-I**

Bagnold, R.A. (1990) *Sand, Wind, and War: Memoirs of a Desert Explorer.* University of Arizona Press, Tucson, Arizona

Bennett, S.J., Hou, Y. & Atkinson, J.F. (2014) Turbulence suppression by suspended sediment within a geophysical flow. *Environmental Fluid Mechanics* **14**, 771–794

Irmay, S. (1960) Accelerations and mean trajectories in turbulent channel flow. *Transactions of the American Society of Mechanical Engineers* **82**, 961–972.

Falco, R.E. (1977) Coherent motions in the outer region of turbulent boundary layers. *Physics of Fluids Supplement* **20**, 124

Leeder, M.R., Gray, T.E. and Alexander, J. (2005) Sediment suspension dynamics and a new criterion for the maintenance of turbulent suspensions. *Sedimentology* **52**, 683–691

Reynolds, O. (1883) An experimental investigation of the circumstances which determine whether the motion of water in parallel channels shall be direct or sinuous and of the law of resistance in parallel channels. *Philosophical Transactions of the Royal Society of London* **174**, 935–982

Reynolds, O. (1895) On the dynamical theory of incompressible viscous fluids and the determination of the criterion. *Philosophical Transactions of the Royal Society of London* **186**, 123–164

Townsend, A.A. (1976) *The Structure of Turbulent Shear Flow*, 2nd edn, Cambridge University Press, Cambridge

Part 5: Climatic Stuff

Chapter 12 Milutin Milanković

Croll, J. (1864) On the physical cause of the change of climate during geological epochs. *Philosophical Magazine* **28**, 121–137

Croll, J. (1867) On the eccentricity of the Earth's orbit, and its physical relations to the glacial epoch. *Philosophical Magazine* **33**, 119–131

Croll, J. (1867) On the change in the obliquity of the ecliptic, its influence on the climate of the polar regions and on the level of the sea. *Philosophical Magazine* **33**, 426–445

Croll, J. (1893) *Climate and Time in their Geological Relations: A Theory of Secular Changes of the Earth's Climate.* D. Appleton and Company, New York

Köppen, W. and Wegener, A. (1924) *Die Klimate der geologischen Vorzeit.* Berlin. English translation and facsimile (2015) *The Climates of the Geological Past.* Borntraeger Science Publishers, transl. B. Oelkers

Chapter 13 Harold Clayton Urey

Cohen, K.P., Runcorn, S.K., Suess, H.E. and Thode, H.G. (1983) Harold Clayton Urey, 29 April 1893–5 January 1981. *Biographical Memoirs of Fellows of the Royal Society* **29**, 622–659

Miller, S.L. (1953) A production of amino acids under possible primitive earth conditions. *Science* **117**, 528–529

Urey, H.C. (1947) The thermodynamic properties of isotopic substances. *Journal of the Royal Society of Chemistry*, 562–581. https://doi.org/10.1039/JR9470000562

Urey, H.C. (1948) Oxygen isotopes in nature and in the laboratory. *Science* **108**, 489–496

Urey, H.C. (1952) On the early chemical history of the Earth and the origin of life. *Proceedings of the National Academy of Sciences* **38**, 351–363

Urey, H.C., Epstein, S., Buchbaum, R. and Lowenstam, H.A. (1953) Revised carbonate-water isotope temperature scale. *Bulletin of the Geological Society of America* **64**, 1315–1326

Urey, H.C., Lowenstam, H.C., Epstein, S. and McKinney, C.R. (1961) Measurement of paleotemperatures and temperatures of the Upper Cretaceous of England, Denmark, and the South-eastern United States. *Bulletin of the Geological Society of America* **62**, 399–416

Part 6: Coda: Legacies and Connections

Chapter 14 Mohos exposed

Bailey, E.B. (1936) Sedimentation in relation to tectonics. *Bulletin of the Geological Society of America* **47**, 1713–1725

Bailey, Sir E.B. and McCallien, W.J. (1950) The Ankara Melange and the Anatolian Thrust. *Nature* **166**, 938–940

Bailey, E.B. and and McCallien, W.J. (1960) Some aspects of the Steinmann trinity, mainly chemical. *Quarterly Journal of the Geological Society of London* **116**, 365–395

Bernoulli, D., Manatschal, G. Desmurs, L. and Müntener, O. (2003) Where did Gustav Steinmann see the trinity? Back to the roots of an Alpine ophiolite concept. *Geological Society of America Special Paper* **373**, 93–110

Gass, I.G. and Masson-Smith, D. (1963) The geology and gravity anomalies of the Troodos Massif, Cyprus. *Philosophical Transactions of the Royal Society of London.* A**255**, 417–467

Gass, I.G. (1968) Is the Troodos Massif of Cyprus a fragment of Mesozoic ocean floor? *Nature* **220**, 39–42

Mitchell, A.H. and Reading, H.G. (1969) Continental margins, geosynclines and ocean floor spreading. *Journal of Geology* **77**, 629–646

Chapter 15 Slip-sliding away

Forsyth, D., & Uyeda, S. (1975) On the relative importance of the driving forces of plate motion. *Geophysical Journal International* **43**(1), 163–200

Rychert, C.A., & Shearer, P.M. (2009) A global view of the lithosphere–asthenosphere boundary. *Science* **324**, 495–498

Rychert, C.A. and 9 others. (2012) Volcanism in the Afar Rift sustained by decompression melting with minimal plume influence. *Nature Geoscience* **5**, 406–409

Rychert, C. A., Harmon, N., Constable, S., & Wang, S. 2020. The nature of the lithosphere-asthenosphere boundary. *Journal of Geophysical Research: Solid Earth* **125**, e2018JB016463. https://doi.org/10.1029/2018JB016463

Schmerr, N. (2012) The Gutenberg discontinuity: melt at the lithosphere–asthenosphere boundary. *Science* **335**, 1480–1483

Schmidt, G.W. (1973) Interstitial water composition and geochemistry of deep Gulf Coast shales and sandstones. *Bulletin of the American Association of Petroleum Geologists* **57**, 321–331

Shearer, P.M. (1991) Constraints on upper mantle discontinuities from observations of long-period reflected and converted phases. *Journal of Geophysical Research* **96**, 18, 147–182

Chapter 16 Unzipping the oceans

Briden J.C., Morris, W.A. and Piper, J.D.A. (1973) Palaeomagnetic studies in the British Caledonides – VI Regional and Global Implications. *Geophysical Journal of the Royal Astronomical Society* 34, 107–134

Dewey, J.F. (1969) Evolution of the Appalachian/Caledonian Orogen. *Nature* **222**, 124–129

Girdler, R.W. (1958) The relationship of the Red Sea to the East African Rift System. *Quarterly Journal of the Geological Society of London* **114**, 79–105

Girdler, R.W. (1962) Initiation of continental drift. *Nature* **194**, 521–524

Fitton, J.G. and Hughes, D.J. (1970) Volcanism and plate tectonics in the British Ordovician. *Earth and Planetary Science Letters* **8**, 223–228

Hess, H.H. (1954) Geological hypotheses and the Earth's crust under the oceans. In: *A Discussion on the Floor of the Atlantic Ocean, Proceedings of the Royal Society of London*, A**222**, 341–348

Hess, H.H. (1962) History of ocean basins. In: *Petrologic Studies: A Volume to Honor A.F. Buddington*, A.E.J. Engel, Harold L. James, and B.F. Leonard (eds), Geological Society of America, 599–620

James, H.L. (1973) *Harry Hammond Hess 1906-1969: A Biographical Memoir*. National Academy of Sciences, Washington D.C.

McKenzie, D.P. and Parker, R.L. (1967) The North Pacific: an example of tectonics on a sphere. *Nature* **216**, 1276–1280

Vine, F.J., and Matthews, D.H. (1963) Magnetic anomalies over oceanic ridges. *Nature* **199**, 947–949

Vine, F. J. (1966) Spreading of the ocean floor: new evidence. *Science* **154**, 1405–1415

Wilson, J.T. (1962) Cabot Fault, an Appalachian equivalent of the San Andreas and Great Glen Faults and some implications for continental displacement. *Nature* **195**, 135–138

Wilson, J.T. (1965) A new class of faults and their bearing on continental drift. *Nature* **207**, 343–347

Wilson, J. T. (1966) Did the Atlantic close and then re-open? *Nature* **211**, 676–681

Chapter 17 Carbon, mountains and cooling

Arrhenius, S. (1896) On the influence of carbonic acid in the air upon the temperature of the ground. *Philosophical Magazine and Journal of Science* **41**, 237–276

Babbage, C. (1835) *On the Economy of Machinery and Manufactures.* John Murray, London

Berner, R.A. (2004) *The Phanerozoic Carbon Cycle.* Oxford University Press, Oxford

Berner, R.A. (2012) Jacques-Joseph Ébelmen, the founder of earth system science. *Comptes Rendus Geoscience* **344**, 544–548

Ébelmen, J-J. (1845) Sur les produits de la décomposition des espèces minérales de la famile des silicates. *Annales des Mines* **7**, 3–66

Lovelock, J.E. (1979) *Gaia: A New Look at Life on Earth.* Oxford University Press, Oxford

Raymo, M.E. & Ruddiman, W.F. (1992) Tectonic forcing of late Cenozoic climate. *Nature* **359**, 117–122

Robinson, J.E. (1991) Phanerozoic atmospheric reconstructions: a terrestrial perspective. *Palaeogeography, Palaeoclimatology, Palaeoecology* **97**, 51–62

Tyndall, J. (1859) On the transmission of heat of different qualities through gases of different kinds. *Proceedings of the Royal Institution* **3**, 155–158

Chapter 18 Isotopes, ice and orbitals

Emiliani, C. (1955) Pleistocene Temperatures. *Journal of Geology* **63**, 538–557

Hays, J.D., Imbrie, J. and Shackleton, N.J. (1976) Variations in the Earth's orbit: pacemaker of the Ice Ages. *Science* **194**, 1121–1132

Imbrie, J. and Imbrie, K.P. (1979) *Ice Ages: Solving the Mystery.* Enslow Publishers, USA

McCave, I.N. and Elderfield, H. (2011) Sir Nicholas John Shackleton 23 June 1937–24 January 2006. *Biographical Memoirs of Fellows of the Royal Society* **57**, 435–462

Olausson, E. (1965) Evidence of climatic changes in North Atlantic deep-sea cores, with remarks on isotopic paleotemperature analysis. *Progress in Oceanography* **3**, 221–252

Shackleton, N.J. (1967) Oxygen isotope analyses and Pleistocene temperatures re-assessed. *Nature* **215**, 15–17

Shackleton, N.J. and Opdyke, N.D. (1973) Oxygen isotope and palaeomagnetic stratigraphy of equatorial Pacific core V28-238: Oxygen isotope temperatures and ice volumes on a 10^5 year and 10^6 year scale. *Quaternary Research* **3**, 39–55

Index

Page numbers in *italic* refer to figures